D1439083

THE
PERMISSIVE
UNIVERSE

The
Permissive
Universe

by
Kirtley F. Mather

UNIVERSITY OF NEW MEXICO PRESS
Albuquerque

Library of Congress Cataloging in Publication Data

Mather, Kirtley F. (Kirtley Fletcher), 1888–
    The permissive universe.

    Includes index.
    1. Religion and science—1946–   I. Title.
BL240.2.M397   1986   261.5'5   85–24618
ISBN 0–8263–0856–2

In memory of
Marie Porter Mather
1889–1971

Contents

Foreword

THE BREADTH OF A MAN'S VISION and accomplishments may best be assessed by the density of odd encounters in unexpected places. One evening, while crossing the Canadian border at a lonely checkpoint in Calais, Maine, I fell into conversation with the border guard who, upon learning that I taught at Harvard, asked: "Do you know Kirtley Mather? He taught me geology in the 1930's. Finest teacher I ever had." A few years later, in the midst of faculty turmoil during the VietNam war, I was reading through the archives of past meetings, seeking verbal ammunition for a speech. I began leafing through records of a past period of contention—those frightening times at the height of McCarthyism when so many professors compromised their principles and fired or curtailed the freedom of colleagues with unpopular views. One man stood out in the reports for his uncompromising defense of academic freedom during its greatest peril. One voice spoke more eloquently and more persistently than all others: that of Kirtley Mather.

I met Kirtley Mather through his role in another dark moment of American history—his defense of science and liberal learning as an expert witness called by Clarence

Darrow to support John Scopes at the famous Tennessee "monkey trial" of 1925. Kirtley, in his early 80's, and still capable of weaving circles around the finest orators at Harvard, gave a yearly guest lecture to my class in the history of life. His talk, recalling his experiences at the Scopes trial, changed little from year to year. At first, I viewed it as a charming evocation of times gone by. Later, as the creationist movement began to surge again, I deemed it mildly related to current affairs, still later as a vital statement about pressing realities and, finally, as a disquisition on immediate dangers.

I loved Kirtley Mather. The half century difference in our ages didn't matter at all. He approached the world with the wonder and excitement of a child, and he treated ideas with the freshness of an earnest undergraduate. He did not dwell upon the past, though he used the richness of his personal experience to inform the present and offer wise advice for the future.

Kirtley Mather was one of those rare men "of enlarged curiosity" (a description applied by Josiah Wedgewood to Charles Darwin) who grace our planet from time to time and who try to integrate the many compartments of our too-specialized world into a coherent vision of life worth living. He was, first of all, a good geologist (he even wrote his Ph.D. dissertation in my own field of paleontology). His perspective from this most integrative of all sciences then expanded to encompass both the factual and moral sides of human life. He struggled all his days to integrate what science taught about the world's structure and history with what religion had learned about our spiritual needs. (Indeed, Darrow called him as a witness to demonstrate that evolutionists need not be narrow minded, hard hearted, atheistic materialists).

Upon his death, at age 90, after such a long and fruitful life, Kirtley Mather left behind a final testimony to his lifelong quest for integration—his last book, *The Permissive*

Universe. Its organizing metaphor—Mather's concept of a universe that permits us enormous freedom to sin or shine, but that imposes constraint by virtue of its own order (that is, permissive, but not chaotic)—stands as an apt and fitting summation of Mather's vision. It resolves the quintessential conflict between order and chaos that characterizes the history of our quest to understand a complicated world that we did not enter by clear design. It also offers hope as we rush into an uncertain future.

Harvard University   STEPHEN JAY GOULD
June, 1985

Preface

AS A LIFE-LONG HABITUÉ of academic halls, I am particularly concerned about the contemporary reexamination of educational programs with special reference to their relevance to life as it really is. It seems to me that the most relevant part of anybody's education—regardless of the height to which he or she may have climbed in his or her formal schooling—is that which enables the person to develop a viable philosophy of life, encompassing a valid conviction concerning his relation to other human beings and a plausible commitment concerning his place and role in the universe as a whole. Then, when I recall that as a geologist I have joyfully spent more than threescore years studying the history of the earth and its inhabitants, I am impelled irresistibly to share with others some of the significant facts and important inferences gained from that activity.

The audience I would like to address is broadly inclusive and widely diversified. Therefore I avoid the use, so far as possible, of the technical jargon characteristic of the discourse within the highly specialized "disciplines" into which the intellectual enterprise of our day is necessarily fragmented. Wherever technical terms must be used for precision and clarity, they are defined, or the concepts they symbolize are described, in the vocabulary that is—or ought to be—the common property of every high-school graduate. Thus the specifications for membership in my audience are contemplated with respect to interest and concern rather than familiarity with the esoterica of science, philosophy, and religion.

Accordingly I start with a prolegomenon comprising a terse summary of one of my major theses: we are living in a universe characterized not only by firm regulations and rigorous guidelines but also by challenging permissiveness. I focus attention on the causes of today's tense situation in human affairs and reach the conclusion that the most fundamental of these causes are the unprecedented increase of knowledge during recent decades and the uses to which the new knowledge has been put. "Explosion of knowledge" is a melodramatic but potentially misleading expression; the proliferation of authentic knowledge is inherently an orderly process. Therefore, as a prime example of that process and as an illustration of the influence of contemporary politics upon it, a brief account is given concerning the sequence of events that led to the discovery of ways and means whereby nuclear energy could be made available for human use.

There are many kinds of knowledge and many proposals for definitions and classifications of the various kinds. I recognize two kinds of knowledge: (1) knowledge about the quantitative (and therefore at least potentially measurable) characteristics and transformations of matter and energy; and (2) knowledge of the qualitative factors subject to evaluation but not to physical measurement in human nature and in the environment in which life has developed. Whether or not the reader agrees with me that these kinds of knowledge may appropriately be designated as scientific and spiritual is unimportant: "a rose by any other name would smell as sweet."

Knowledge concerning the past is essential to any helpful understanding of the present and is prerequisite to any intelligent planning for the future. Indeed it is implicit in the term, reexamination, that I have just used. To the geologist, history stretches far back beyond the advent of man with his written records and eloquent artifacts. Therefore I summarize the record of geologic life development and draw inferences from that record concerning the strategy and tactics of the broadly inclusive process of organic evolution. Here attention is paid not only to so-called "laws of evolution" but also to the permissiveness that seems to be present within the limits set by those regulations and directives. The emergence of mankind along the path of life during geologic time is then sketched. From the factual record it is apparent that the spirit of cooperation and the flair for organization were of prime importance in the sequential stages of human evolution. Presumably they will be at least as influential in the new evolutionary stage in which we find ourselves today.

The more successful the scientific enterprise the more certain are we that we are living in a universe of law and order. Such a universe must have some kind of administration. The prosperity, indeed the very survival, of mankind depends upon the ability of men and women to order their ways of life in accordance with the administrative guidelines. It is necessary to inquire concerning the nature and functions of that administration. I venture the suggestion that there may be a field of spiritual force operating administratively in a manner analogous to that of the gravitational, electromagnetic, and other fields of force now recognized in the physical sciences. Whether or not that outrageous hypothesis is worthy of credibility, there is no question but that men are motivated by spiritual aspirations as well as by material objectives; they are aware of esthetic and ethical values that transcend physical realities. Value judgments have their roots far back in geologic history; they will be decisive concerning man's future.

I have confidence, despite the gloomy prognostications

of some of my more pessimistic colleagues, that mankind does have a future. Surveying the possibilities of that future, it becomes apparent that its nature will be determined more by the resources of the human mind and spirit than by the physical resources of the terrestrial environment. The intellectual resources now available seem adequate to solve the many obdurate problems casting dark shadows on the human situation, but it is not so apparent that the spiritual resources are competent to meet the challenge of our troubled time. I therefore consider the role of religion in the new world of science and technology. This forms two parts, each of which is an introductory study of its subject rather than a full-blown treatise on it. First, I suggest the tenets and objectives of a religion in general that seems to me to be compatible with the world view of modern science. Second, I look specifically at the Christian religion and present my personal view about its relevance to human affairs during this last third of the twentieth century. When the husks of traditional dogmas and creeds, most of them inherited from the Middle Ages, are stripped away, there is disclosed a spiritual directive and power, reflected in the life and teachings of Jesus Christ, that could save mankind from self-destruction.

Even as I started with a prolegomenon, so I end with an epilogue comprising a terse summary of another of my major theses: at the moment *Homo sapiens* has a unique opportunity—unique in time, in space, and in comparison with all other species of animal—to move forward from his present stage of evolutionary progress to new ways of life that will ensure his enduring welfare, individually as well as collectively. That glorious future is not, however, guaranteed for him; it is only an opportunity which may be seized or rejected, as men and women determine.

Acknowledgments

IN UNDERTAKING the perilous task of expressing in words my convictions and speculations concerning matters so broad in scope and profound in depth, I am well aware of my great indebtedness to others. Ideas are light baggage on the journey through life; they are easily transferred from one traveller to another, subconsciously as well as consciously. Among those to whom I am most deeply grateful are the authors and editors of the books and articles listed in the footnotes at the end of each chapter, most of whom are long-time friends and co-workers in various enterprises stimulated by goodwill as well as intellectual curiosity, but there are many others who must remain nameless. At the risk of neglecting some to whom my debt is equally great, I put here a word of special thanks to the late F. L. Kunz, key executive of the Foundation for Integrated Education, now known as the Center for Integrative Education, Henry Margenau, professor of physics and natural philosophy at Yale University, Ralph W. Burhoe, honorary president of the Institute on Religion in an Age of Science, and the late Harlow Shapley, astronomer-statesman-humanist extraordinary. Needless to say, full responsibility is mine alone for any inaccurate state-

ment of fact or erroneous deduction or any scatter-brained
ideas that may inadvertently or allegedly appear in these
pages.

Albuquerque, New Mexico   KIRTLEY F. MATHER
October, 1975

PUBLISHER'S NOTE:

Dr. Kirtley Fletcher Mather's last book "The Permissive Universe" is published posthumously through the editorial and financial efforts of Dr. Sherman A. Wengerd, Professor of Geology,
Emeritus, The University of New Mexico. Further credit is due
the University Faculty Publications Committee, and financial
support provided in part through the "Mather Fund" of The
University of New Mexico Foundation established by Dr. Mather's family: Jean Mather Seibel, Julia Mather Seils, Dr. Leroy C.
Seils, Florence Mather Wengerd, and Dr. Sherman A. Wengerd.

Prolegomenon

BELIEFS ABOUT THE NATURE of the world in which men live and ideas concerning man's place in the universe have changed drastically through the centuries. Expansion of knowledge gained through scientific research makes it necessary to ask the big questions over and over again. Appraising and reappraising the intellectual respectability of philosophical concepts and theological doctrines may seem an irksome task, but it cannot be evaded.

The world view of modern science is notably different from that presented only a few decades ago. The part of the universe which is studied in the physical sciences continues to appear completely neutral toward mankind. Its operations neither favor the life of mankind nor oppose it; they merely provide a stage on which the drama of life may be played. The stage has limits and it is not inert, but ants and tigers can play their roles on it as commendably for them as can man or any other creature, each in its own way. The operations of the physical universe are not, however, so completely mechanical as they appeared to be in the days before nuclear physicists discovered secrets previously concealed within the atom. The analogy of a gigantic, ultra-complicated machine is no longer as appropriate as it seemed a half-century ago. Nevertheless, there is no evidence that the physical universe was designed especially for mankind or for human beings of any particular race, nationality, political system, or religious creed.

Much of the part of the universe which is studied in the biological sciences likewise appears to be neutral toward mankind. The regulatory principles pertaining to metabolism, respiration, sense perception, and the like are identical for man and other animals. The processes of natural selection, operative in our part of the universe, have been said to favor mankind. How else could his eminence in the organic world be accounted for? But the same can be said for many others among the living species of animals and plants. Even so, there may be something in the universe that has favored the particular intellectual abilities displayed by man and thus may account for his actual or potential supremacy over all other creatures.

The part of the universe which is studied in the social sciences presents a somewhat different picture. This may be due to the fact that the laws of nature in this area are not so well understood and formulated as they are in other areas. Or it may be that man himself is administratively responsible in this enclave within the over-arching administrative domain of nature. It is here that students of religion find some of their deep concerns. To what extent, if any, does the universe favor men of integrity who seek justice and love mercy over those who lack these spiritual qualifications?

In the present state of knowledge it is unrealistic to expect unanimous agreement concerning the relations between mankind and the universe as a whole. Some may be convinced that it is a friendly universe; although certain of its aspects are neutral toward human beings, others display a benign or even paternal solicitude for human welfare. If so, there is little for good people to worry about. Others may be convinced that it is a hostile universe, fun-

recent years. These are the unprecedented rate of increase in human population, the prevalence of rising expectations, and the extraordinary expansion of knowledge. When any one of these winds of change blows as violently as each is blowing today, there is trouble for us all. When all three hit us simultaneously, the resulting whirlwind produces a crisis such as mankind has never before encountered.

Of these three troublesome factors in our lives, the expansion of knowledge is the most basic. The unprecedented increase in the rate of population growth is essentially a result of the triumphs of modern medical science and public hygiene. The average life of human beings has been greatly prolonged and death rates significantly reduced by the application of new knowledge about disease and health, but there has been no compensating reduction in birth rates. Only a small fraction of the human race has thus far enjoyed all the productive efficiency and comfortable existence made possible by modern technology, though the great majority of the underprivileged have learned about the new ways of life through the marvelously efficient channels of communication now available. The demand for opportunities to share the benefits accruing from the applications of scientific knowledge for human betterment arises in every land. It comes from the highly industrialized nations with their enclaves of poor as well as from the numerous underdeveloped nations of the earth. The extraordinary increase of knowledge during the last few decades is responsible for what we appropriately may call the revolution of rising expectations. New knowledge and its use in human affairs may be the most fundamental cause of our time of trouble.

To change the figure of speech and express this thought in another way, we are suddenly living in a world we had not known before. We arrived in this new world not as passengers on a Mayflower but by way of scientific lab-

oratories and mechanized industries. The image of a
mushroom cloud rising over a city in Japan, an island in
the West Pacific, or a testing range in Nevada or Siberia
persists in the background of our minds. Millions of us
watch a football game in the Pasadena Rose Bowl as it
appears on television screens in our cozy living rooms,
hundreds or thousands of miles away. Perhaps even more
of us watch a President of the United States as he talks to
us from the White House. Electricity produced from nuclear energy pulses through transmission lines in Pennsylvania, in England, and in Moscow, and by the time
these words are in print in many other localities as well.
Man-made satellites circle the earth in about the same time
that it takes us to drive across any of our larger cities.
Electronic computers identify each of us by a number and
check our honesty in the calculation of our taxes.

In all probability the most impressive demonstration of
the newness of our new world took place in July of 1969
when upwards of a hundred million human beings, the
world around, watched astronaut Neil Armstrong on their
television screens as he descended from his lunar module
to the surface of the moon and heard him say "That's one
small step for a man, one giant leap for mankind." The
more one contemplates what was involved in that event,
the more awesome it seems. Scientific knowledge, insatiable ingenuity, almost incredible mobilization of human
and terrestrial resources, and man's unique capacity for
organization had made possible not only the journey to
the moon but also the opportunity for all of us to observe
simultaneously the arrival at its goal. Nothing that has
occurred to date could drive home more effectively the
difference between the world in which we now live and
the world into which we adults were born.

No wonder the watchmen in their ivy-clad towers tell
us that the atomic revolution is destined to exert an even
greater impact upon human life than did the industrial

revolution that began nearly two centuries ago. As I see it, however, neither of these is a revolution in the real meaning of that word. Each is in fact a stage in the ongoing evolution of man and society, an evolution that began a few hundred thousand years ago and will continue as long as mankind succeeds in maintaining human life on the face of the earth. The pace of evolutionary change has at times been slow, at other times rapid. Right now, it is extraordinarily swift. In the middle third of the twentieth century, the ways of life were changed more completely than they were changed during the entire preceding century and a half, even though that period encompassed the industrial revolution, hence the illusion that we have been catapulted into a new age; hence also the fact that ours is a time of trouble.

Probably I should attempt to prove my thesis that the new world is a result of accelerated evolution rather than abrupt revolution. Ask a dozen nuclear physicists to tell you the date that should be put in the history books to mark the beginning of the Atomic Age or, as they more correctly call it, the Age of Nuclear Energy. You will certainly get two different answers, more likely three, and possibly even six or eight. The most favored dates are December 2, 1942, and July 16, 1945. But some of the experts say, late in 1938, and a few can be found who would pick 1932 or even 1896.

The record of this exploration and its discoveries is not only significant for the historians of science; it can tell all of us much about the nature and procedures of the scientific enterprise. It is a prime example of the response of men of science to the challenge of the unknown and their motivation as they devote themselves to research along the endless frontier of expanding knowledge. I will comment on some of these human aspects of the technologic enterprise after I have sketched the factual record.

The fact that certain kinds of matter spontaneously discharge energy was discovered in 1896. A French scientist, Henri Becquerel,[1] had left an unopened packet of fresh photographic plates near some crystals of a uranium-bearing compound, the fluorescence of which he was investigating. When the plates were later developed, he found to his surprise that they had been darkened as if by exposure to light. Seeking a rational explanation for this unexpected phenomenon, he placed other sealed packets of unexposed photographic plates near a variety of fluorescent minerals and found that only those minerals containing uranium compounds produced a similar result. On the second of March of that year (1896), he reported to the French Academy of Science that uranium-bearing crystals and ore fragments emitted some kind of radiant energy, akin to light in its effect on photographic emulsions, but far more energetic, inasmuch as it penetrated through opaque materials. A little later, Pierre and Marie Curie[2] called this energy radioactivity and identified the most highly radioactive substance in the uranium ore as a different element from uranium. This they appropriately named radium. They also distinguished three kinds of radiation from that element, differing in their ability to penetrate opaque screens of varying thickness and diverse materials. These they designated as alpha, beta, and gamma rays.

In 1905, Albert Einstein[3] made the audacious suggestion that radioactivity is the gradual transformation of matter into energy. Only a man of great courage and greater imagination could have proposed so unorthodox an idea at a time when most if not all scientists believed firmly that the atoms, of which all matter is composed, are indestructible. In a very real sense, Einstein's proposition was an intuitive perception, a scientific hunch. There was no way at that time to prove or disprove its validity. In 1910, when Sir Ernest Rutherford[4] disintegrated an atom of hydrogen by bombarding it with rays emitted by radium, there was no evidence that energy had been released. In 1932, however, Rutherford's compatriot, John Cockcroft,[5] demonstrated the fact that the destruction of atoms actually did release energy. Attention began to be paid to Einstein's suggestion and to his further proposition that the transformation of matter into energy would be in accordance with the equation, $E = mc^2$ (the amount of energy produced equals the mass of the transformed matter multiplied by the square of a constant). If, as he thought, the c in that equation represented the velocity of light, such a transformation would involve the release of energy of a completely different order of magnitude from that produced by chemical reactions during the combustion of coal or oil. Investigations of atomic structure and particularly of the composition of atomic nuclei might after all have repercussions of far-reaching significance in the world of practical affairs beyond the walls of scientific laboratories.

It was necessary to learn much more about the nature and behavior of the components of atoms before that possibility could be recognized. One of the most portentous events in the sequence of discoveries was the recognition of the neutron among these components. It was relatively easy to identify the electron and the proton among subatomic entities. Electrons carry negative charges of electricity, protons positive charges; the name *electron* had been in the vocabulary of physicists and chemists since 1891 and the name *proton* since 1914. For more than a decade prior to 1932, physicists had believed that there must be a third subatomic particle of about the same mass as the proton but displaying no net electrical charge. Indeed the name *neutron* was given to it by W. D. Harkins[6] in 1920, hypothetical though it was at that time. Its reality was demonstrated in 1932 by James Chadwick,[7] not by direct observation but by the results of experiments which could be explained only on the assumption that the radiation he had induced with beryllium consisted of a stream of par-

story has since been told many times.[15] Thanks to a truly extraordinary mobilization of procurement and processing agencies, an unprecedented quantity of pure uranium, moderating minerals, and testing devices had been assembled in what had been a squash court beneath the West Stand of Stagg Field at the University of Chicago. Carefully the last blocks of uranium and carbon were added to the pile, and the control rods were slowly drawn out. The hopes of Enrico Fermi[16] and his co-workers were completely fulfilled; the uranium pile worked. For the first time in history, a self-sustaining chain reaction of nuclear fission released useful quantities of energy under human control.

That absolutely novel achievement might seem to be an appropriate milestone to mark man's entry into the Age of Nuclear Energy. But the energy released that day was never used. Those who prefer to count the years of the new age from July 16, 1945, have at least a talking point. The scene was utterly different. Just before dawn, an A-bomb was detonated in the desert near Alamogordo, New Mexico. Watching from distances of ten miles or more, the few observers of that spectacular event saw the burst of light, more blinding than the sun, and the many-colored mushroom-shaped cloud boiling upward to a height of seven or more miles. They knew then that their theories of sub-atomic processes were correct, that their measurements of nuclear components had been accurate. Einstein's insight was valid, his equation was true.

That first A-bomb was only a test bomb. It vaporized the steel tower on which it had been placed and melted the desert sand around the tower's base. The blast of its shock wave was wasted on the desert air and its lethal radioactivity was hoisted harmlessly into the stratosphere. Shameful though it may seem to many, the first actual use of nuclear energy was made when the A-bombs were exploded over Hiroshima on August 5, 1945, and over Nagasaki three days later. No one, however, has suggested that we should date the beginning of the Age of Nuclear Energy from either of those ghastly days.

This may not be so much a tacit expression of guilt as an implicit recognition of the fact that the use of nuclear energy in weapons of warfare is only one of many highly specialized applications in human activities of that newly available kind of energy. Power may be derived to turn the wheels of industry and commerce or to light and heat the homes and offices of coming generations. Radioactive isotopes, artificially produced in nuclear reactors, have widely diversified uses in many industries. In the march of progress toward current achievements in all these and other beneficial applications of nuclear energy, December 2, 1942, is the significant date.

Far more important than the sequential dates in that record are the subjective inferences that may be drawn from it. This particular segment of the scientific enterprise was impressively a collective operation involving many individuals of several different nationalities and racial heritages. Each of them used information gained by the experiments and observations of their predecessors. Until late in 1938, the ideas and discoveries of each were freely shared with all their co-workers anywhere in the world through conversations, correspondence, and widely available publications; freedom to communicate with others was the life-blood of the entire enterprise. The news about the sudden destruction of Hiroshima and Nagasaki in August 1945, and the nature of the weapon used there, came as an almost traumatic shock to most of the literate people the world around. To many, it seemed the result of a revolutionary advance in knowledge about the structure of atoms. But to tens of thousands of physicists and chemists in a score of countries that news was no great surprise. They had been aware of the step-by-step evolution of that kind of knowledge between 1896 and 1938, and they could easily project that evolution a couple steps farther, beyond the curtain of secrecy dropped in late 1939. It is almost certain that, even without the stimulus of the desires for an ultimate weapon, energy from the chain reaction of nuclear fission would have become available by the early 1950s. The war only accelerated by a few years the evolution of scientific knowledge prerequisite to the construction of nuclear reactors.

Until 1938, each of the researchers was motivated by an intense desire to penetrate beyond the known into the beckoning area of the unknown; it was an outstanding example of the quest for knowledge for its own pure sake. There may have been faint whispers of a desire to gain prestige, especially among one's peers, and to escalate one's status and salary, especially in academic institutions, but nobody gave much thought to the usefulness, or lack thereof, of the specific items of new knowledge stemming from their experiments and observations.

Then came a profoundly significant change. The discoveries in 1938 and 1939 that nuclear fission could actually be induced in uranium atoms and that the regulations governing the behavior of subatomic entities permitted chain reactions capable of releasing vast amounts of energy gave the nuclear scientists a new motive. That the knowledge they were gaining could be useful in the practical affairs of mundane life was no longer a vague hope. Henceforth, nuclear research was to be pursued for the contribution it might make to the efficiency with which men did the things they wanted to do. This scientific enterprise, like so many others, moved out of its academic cloister to become a servant of mankind. At the same time the sociopolitical climate, by which scientists are always affected, was rapidly changing. The world was sliding relentlessly into the abyss of World War II. The men responsible for the destiny of the most powerful nations wanted ever more powerful weapons of mass destruction. There was much talk about an ultimate weapon, and the newly available nuclear energy might possibly fulfill that desire. The nuclear physicists of the United States and Great Britain, including many of the most talented who had recently immigrated from Germany and Italy to escape the tyranny already manifest in their homelands, were persuaded that the demonic forces under Hitler's rigid control were embarking on the task of producing atomic bombs. There seemed no alternative; they must complete the application of the new knowledge to the development of new weapons before the enemy could use it in that way. The scientific community responded as always to the social and political pressures of the time and place.

As a consequence, one of the most pernicious of the troubles of our time is the presence in national arsenals of nuclear weapons which if used could conceivably destroy all human civilization. The arms race among the great powers, and all that it means in the life of mankind, stems in large part from the rapid evolution of knowledge about sub-atomic entities and activities, a part of which I have just sketched. It should be stated quickly, however, that it is not this new knowledge *per se* that should be blamed for such dire consequences; it is the uses to which that knowledge has been put. It has already been used for many beneficial purposes, some of which I will call attention to in subsequent pages, and there are more to come. Moreover, in justice to the scientists and technologists who produced those first nuclear devices in the Los Alamos Laboratories, it should be noted that several of them made strenuous, although futile, efforts to deter the dropping of those bombs without warning on Japanese cities during that fateful August of 1945.[17]

As a matter of fact, the daily activities of all but a small minority of the earth's inhabitants have not as yet been noticeably affected by the availability of nuclear energy. Mankind is still in process of adjustment to conditions arising from the application of scientific knowledge gained by research quite other than that of nuclear scientists. Evolution toward a technological culture started slowly in many countries soon after the invention of the steam engine in the latter part of the eighteenth century. It gathered momentum throughout the next hundred years. In the twentieth century, it has gained such speed that few of us have yet become mentally, emotionally, and spiritually adjusted to the new environment in which we will spend the rest of our lives.

Consider first some of the changes in the American way of life that have resulted from the impact of science and technology during the first two-thirds of the twentieth century. Thanks to power-driven machinery, the average American worker produced nearly three times as much per hour in 1957 as did his predecessor in 1907. Whereas the average number of hours of employment per week was 60 in 1900, it was less than 40 in 1960. More than 97 percent of all American homes, rural as well as urban, are today supplied with electricity. More than 80 percent of them have mechanical refrigerators and nearly as many have gas or electric cooking equipment. With the leisure hours that all these things bring, it is not surprising that nearly one fourth of the total expenditures of the American people is now for recreational goods and services.

Anyone can easily compile a long list of things he has or does in his daily life that his parents did not have or could not do when they were the age that he is now. Most of such lists would include watching TV programs, going to an air-conditioned theatre to see an audio-movie in color, buying frozen foods in a supermarket, giving blood to the Red Cross, and getting a prescription filled for a sulfa drug or an antibiotic. Some would include flying in a regularly scheduled airplane from New York to Los Angeles in less than five hours or telephoning a relative in Australia by ways of an orbiting satellite space switchboard.

No matter how long the list of such contributions of

science and technology to modern life, it would never be complete. One of the most startling facts about the new world in which we live is that it is a world in the making. Each year, almost each day, will continue to bring something new and different. With man-made satellites and planetoids circling through outer space, reaching for the moon began to lose its aura of impossibility, and now that Apollo astronauts have left their footprints on the lunar landscape that expression is no more than a quaint echo from an outmoded past.

Nor does any such list give an adequate picture of what is happening to us in these years of our lives. Science and technology have drastically changed the relations between men and their environment and will continue to change those relations still more in the future. More importantly, science and technology are changing the relations between man and his fellowmen in less obvious but actually more significant ways.

All people who enjoy the efficiency, convenience, and comfort of life in the age of nuclear energy are profoundly and inevitably interdependent. That is a major reason why we are living in a time of trouble. Every item on every list of distinctive contributions of science and technology to late-twentieth-century life is available because many persons whom the user never sees are busily at work somewhere in the background. There are the workers in the factories where the many pieces of equipment are made, the miners, drillers, farmers, and fishermen who produce the raw materials. Many others are constantly on the job to keep electricity or gas flowing into our homes, to service our telephones, to supply us with running water. When we travel, whether by automobile, street-car, subway, railroad train, bus, or airplane, countless people are carrying out their diverse responsibilities in order that our journey may be swift and safe. We see only a few of them—the man at the gas pump in the service station, the airline pilot and stewardess, the bus driver, for example—but we know there are many more behind the scenes. We have what we have and do what we do because hundreds or thousands of other people are working for us, and we for them.

It is estimated that the average American has the equivalent of three dozen servants working for him all the time on a 40-hour a week basis. Rare indeed was the feudal prince, the desert sheik or the 19th century financial wizard who had it so good. Gone for us are the days of toil from dawn till dark, of homespun clothing, of butter churned by hand in the kitchen. But gone also is the independence implied by the traditional expression that a man's home is his castle. The physical interdependence of men in our technologic culture goes far beyond anything that John Donne could possibly have imagined in the early part of the seventeenth century when he wrote "No man is an Iland, intire of it selfe; every man is a peece of the Continent, a part of the maine."

This increasing dependence upon each other is in part a result of the need for specialization of functions and skills. Specialization increases with every advance in knowledge gained through scientific research and with the application of that knowledge to the activities of human beings. To coordinate the activities of large numbers of highly skilled individuals, each of whom is a specialist in some field of employment, is a major objective of civilization. Complex structures of economic and social organization have developed in every community in the effort to attain that objective, most of them without much conscious long-range planning.

The reality of our interdependence, one with another, is demonstrated whenever something happens to interfere with our personal activities. A hurricane or tornado may destroy the power lines which bring electricity to our homes; we are immediately aware of our dependence upon the engineers in the power plant and the maintenance men along the line. Heavy snowfall may isolate a town for a few days; food supplies dwindle, streets to the hospital are blocked. We rejoice when roads are cleared and the townspeople are able again to go about their affairs. A strike of the workers in a key industry or transportation service may paralyze community enterprises and seriously interfere with daily routines. Our dependence upon others becomes a matter of real concern.

Mutual dependence is also a result of geographic and geologic factors. Specialization of human occupations is imposed by local variations in natural resources. City dwellers must obtain food produced by farmers, fishermen, and horticulturalists. Farmers must get tools and machinery from factories in industrial towns or cities, many of which were originally located where water power or coal was close at hand. Variations in soil and landscape, combined with diversity of climate, determine the uses to which land may be put. Those who till the ground for wheat or corn in the north-central states must exchange their produce with the citrus-fruit growers of the south, the cattle ranchers of the western plains, and the sugar beet producers of the Rocky Mountain region or the sugarcane growers in sub-tropical lands, if they are to be well-nourished. Malnutrition in the United States, indeed in almost all parts of the world, is usually due to lack of the right kinds of food rather than to lack of enough food.

Textile mills in New England, where waterpower stimulated their construction a century and a half ago, must get cotton from the southern states or import it from overseas, wool from wherever sheep are herded, and in these recent years, synthetic fibers from wherever they are manufactured. Ores of iron, lead, zinc, copper, silver, gold, and other metals must be mined in the relatively small and widely scattered localities where they occur. By no means every state in the Union has oil or gas fields or coal seams. Only a dozen states can produce in large amounts these sources of energy and of petrochemicals.

Whatever may have been the case in past centuries, it is quite certain that in this age of science and technology, no man in America can live to himself alone for very long. Interdependence is like the yoke used in pioneer days to harness oxen to a plow or wagon. At times, especially at first, it seems to restrain, to chafe, and to limit freedom. Yet always it provides the means for more efficient use of natural resources for the welfare of mankind. Sooner or later it stimulates men to develop their capacity for achievement in the fine art of cooperation.

This applies to the political units in which men organize themselves as well as to men as individuals. The townships or counties within a state are interdependent; so also are the separate states within the United States. That is why we are "one nation, indivisible." Our failure, thus far, to provide "equal justice for all" and a fair deal for each within that kind of nation is a major factor in the advent of our time of trouble.

Even as the expansion of knowledge and its technologic applications has deepened the interdependence of individuals and political units within each nation, so also has it increased interdependence among all nations of the earth.

The metallic ores, the nonmetallic deposits of economic value, and the mineral fuels that are basic requirements of modern civilization occur under certain well-defined geologic conditions. Their distribution is by no means haphazard or unpredictable. Certain kinds of rocks contain rich ores of the metals but no petroleum or coal. The latter are found only in certain other kinds of rocks. Knowledge about the origins, geologic relations, and geographic occurrences of the earth's mineral resources has grown rapidly during the twentieth century. Geologists have explored even the most inaccessible localities on all the continents.

Their observations of rocks exposed at the surface of the earth have been amplified by the use of geophysical instruments to obtain information about rocks concealed from view at considerable depths below the surface. Most recently, the scientists who investigate the earth's natural resources have begun to use the newly developing techniques of remote sensing, whereby some of the information they desire is obtained from low-flying satellites (three to five hundred miles up), such as ERTS (Earth Resources Technology Satellite), the first of which was launched by NASA in 1972, and Skylab.[18] Enough is now known to justify the assertion that no nation, with the possible but highly improbable exceptions of the Soviet Union and the People's Republic of China, embraces within its political boundaries a sufficient variety of geologic structures to give it adequate supplies of all the metals necessary for modern industrial operations. Likewise, no nation, again with the same two possible exceptions, enjoys a sufficient variety of climatic conditions to permit all kinds of foodstuffs to be grown on its farms or gathered from its forests or to allow the growth of all the various plants contributing essential raw materials to its industry.

When ranked according to geographic dimensions, the United States of America is third among the nations, following the Soviet Union and the People's Republic of China. Its variety of geologic structures and its fortunate relation to climatic zones make it one of the most favored nations from the point of view of natural resources. Nevertheless, it is completely dependent upon other nations for many essential raw materials of modern industry.

American manufacturers must import nickel, tin, antimony, chromium, and platinum for the fabrication of many important articles. For some of their uses, satisfactory substitutes may be found among our domestic ores but for others this is impossible. Tin is an interesting example. Its ores are scarce in the highly industrialized countries where it is a necessary ingredient of the only alloys that are satisfactory for the bearings of high-speed machinery. There are few ores of tin in North America; the puny deposits of that metal in Europe are sufficient to meet less than 5 percent of the needs of Europeans. The only rich and extensive ores of tin are in Malaya and Bolivia, where at present there is little demand for it. We Americans could get along without tin cans if we had to, but it would be an expensive proposition. Stainless steel or aluminum cans could be used for some things, glass or plastic containers for others. But tin we must have to keep the wheels of industry and commerce whirling.

Similarly, the climate of the United States influences the variety of plants that can be grown within its borders. We are forced to import, either from foreign countries or from our overseas possessions, all the bananas, coffee, tea, camphor, tropical oils, jute, sisal, quinine, and natural rubber that we consume. Synthetic rubber is as good for many purposes as natural rubber; indeed it is far better for a few uses, whereas for others it is far inferior. Other synthetic products are also available, or are likely soon to be available, to replace some of the items on that list. Even so, there is at the present time no escape from the dependence of the United States upon foreign sources for many of the materials of life. Although the specific details and pattern of that dependence will change in future years, it will continue in some form or other for an indefinitely long time to come.

A "have" nation like ours, in which a high standard of living is made possible by rapid consumption of domestic mineral resources, must eventually become a "have not" nation with respect to those resources. Nations which are slower in the exploitation of such resources will then be looked upon as the wealthy possessors of earth's bounty. A prime example of this was demonstrated by the energy crisis of 1973 and 1974, with its shortages of gasoline at

filling stations across the country and its dire pronouncements about the troubles of our time.

We now know that before the first oil well was drilled
near Titusville, Pennsylvania, in 1859, the United States
possessed about 30 percent of the world's total petroleum
resources. Throughout several decades, before the centen-
nial of that historic event was celebrated, the U.S. produced—and consumed—each year more than 60 percent
of the world's total annual production of that energy resource. The consequences were inevitable. In the 1960s the
U.S. was increasingly dependent on importation of foreign
oil to fill the steadily enlarging gap between its consumption and its production. The energy crisis may have been
a sudden and troublesome shock to millions of Americans,
previously unaware of their nation's dependence upon
other nations for energy resources, but it was no surprise
to most geologists. Some of us had given warning, as far
back as the early 1940s,[19] that such a situation might arise
later in the twentieth century.

I will put much more about energy resources for man's
future in chapter 8, but here I must give a word of comfort
for those who are worried about energy in this present
time of trouble. We now know that prior to the first mining
of coal for the newly invented steam engines of the late
eighteenth century, the United States possessed about 30
percent of the world's total coal resources which is the
same percentage figure as for petroleum. Since then the
U.S. has produced—and consumed—each year, on the average, just about 30 percent (not 60 percent, as for petroleum) of the world's total annual production of coal.
Obviously, if this percentage rate continues, the U.S. will
not become a "have not nation" for coal before the rest of
the world also exhausts its coal resources, some several
hundred years from now. For this reason, in its effort to
escape the energy crunch by becoming self-sufficient in
energy production, the U.S. should give high priority to

the consumption of coal instead of petroleum, as well as to the development of nuclear, geothermal, and solar energy. Important also is the curtailment of unnecessary uses of gasoline and electricity and the elimination of waste in their use, of which so many of us are guilty. Even though proper steps are taken along all these lines, and if domestic production of petroleum is significantly increased by extensive drilling in Alaska and on continental shelves offshore, energy self-sufficiency cannot be attained by 1990 or 1995. It certainly has not been achieved by 1975, as some had hoped.

In this case, when expansion of knowledge and its application in practical affairs got us into trouble, further advances in knowledge and its application, combined with wiser use of the knowledge already available, may get us out of that particular predicament. That has happened over and over again in the history of mankind.

NOTES

1. Antoine Henri Becquerel (1852–1908) was professor of physics in the Ecole Polytecnique, Paris, France, when he discovered what were first called "Becquerel Rays" but later became known as radioactivity.

2. Pierre Curie (1858–1906) was a member of the faculty and research staff of the Sorbonne, Paris, France, throughout his professional life. Marie Curie (1867–1934) received her diploma as *Licence és Science Physiques* at the Sorbonne in 1893, married Pierre Curie in 1895, collaborated with him in the study of radioactivity, and succeeded him as Professor of Physics after his death.

3. Albert Einstein (1879–1955) received his Ph.D. degree from the University of Zurich in 1905, was a member of the faculty and research staff of various institutions in Switzerland, Czechoslovakia, and the Netherlands prior to 1914 when he became Director of the Kaiser Wilhelm Institute and Professor of Physics in the University of Berlin. In 1933 he renounced his German citizenship and came to the United States where he was appointed for life as a member of the Institute for Advanced Studies, Princeton, N.J.

4. Sir Ernest Rutherford (1871–1937) was born and educated in New Zealand, worked under J. J. Thomson in Trinity College, Cambridge University, England, from 1895 to 1898, was Macdonald Research Professor of Physics in McGill University, Montreal, Canada, from 1898 to 1907 and Langworthy Professor of Physics in Victoria University, Manchester, England, from 1907 to 1919 before becoming Director of the Cavendish Laboratory at Cambridge, England, in 1919, a post that he occupied until his death in 1937.

5. Sir John Douglas Cockcroft (1897–1967) was at that time a Fellow in St. John's College, Cambridge University, England; later, from 1946 to 1958, he was Director of Britain's Atomic Energy Research Establishment.

6. William Draper Harkins (1873–1951) was at that time Professor of Physical Chemistry in the University of Chicago.

7. Sir James Chadwick (1891–   ) was Lecturer and Assistant Director of Radioactive Research in the Cavendish Laboratory, Cambridge University, England, when he demonstrated the reality of neutrons.

8. Otto Hahn (1879–1968) was Director of the Kaiser Wilhelm Institute in Berlin from 1928 to 1945.

9. Fritz Strassmann (1902–   ) was in charge of the Department of Chemistry in the Kaiser Wilhelm Institute in the late 1930s and early 1940s.

10. Lisa Meitner (1875–1968) was in charge of the Department of Physics in the Kaiser Wilhelm Institute from 1917 to 1938 and thereafter was a member of the staff of the Nobel Institute in Stockholm, Sweden.

11. Otto Robert Frisch (1904–   ) was engaged in research in Copenhagen, Denmark, from 1934 to 1939, in Birmingham and Liverpool, England, from 1939 to 1943, and from 1943 to 1945 participated in the work at Los Alamos, New Mexico, that culminated in the explosion of the test bomb near Alamogordo. Since 1947 he has been a professor in the Cavendish Laboratory in Cambridge University, England.

12. Niels Henrik David Bohr (1885–1962) was Director of the Institute for Theoretical Physics in the University of Copenhagen, Denmark, from 1920 to 1962. He fled to Sweden from the German occupation of his homeland in 1940 and from 1943 to 1945 he served in an advisory capacity in the Los Alamos, New Mexico, Laboratories.

13. Leo Szilard (1898–1964) migrated to the United States from

Hungary in 1937 and conducted research in atomic energy at Columbia University and the University of Chicago until 1946. In 1942 he worked with Enrico Fermi in the development of the first nuclear chain reaction. He was a professor at the University of Chicago from 1946 until his death in 1964, specializing in molecular biology.

14. Walter H. Zinn (1906–   ) came to the United States from Canada in 1930 and was a member of the faculty of Columbia University and of the City College of New York until 1941, when he joined Enrico Fermi's team at the University of Chicago. From 1946 to 1956 he was Director of the Argonne National Laboratory of the Atomic Energy Commission.

15. Among the best of those accounts are: James Phinney Baxter, II, *Scientists Against Time* (Boston: Little, Brown and Company, 1946); paperback edition (Cambridge, MA: M.I.T. Press, 1968). Arthur H. Compton, *Atomic Quest* (New York: Oxford University Press, 1956). Stephane Groueff, *Manhattan Project* (Boston: Little, Brown and Company, 1967). Richard G. Hewlett and Oscar E. Anderson, Jr., *The New World* (University Park, Pa.: Pennsylvania State University Press, 1962).

16. Enrico Fermi (1901–1954) migrated to the United States from Italy in 1939, and was Professor of Physics in Columbia University until 1942 when he became the leader of the research group at the University of Chicago, responsible for the achievement of the first nuclear chain reaction. He remained at Chicago in the Institute for Nuclear Studies until his untimely death in 1954.

17. The endeavors of scientists and engineers to influence the social and political consequences of their success in making nuclear energy available for human use are well reported by Alice Kimball Smith in *A Peril and a Hope* (Chicago: University of Chicago Press, 1965).

18. Lloyd Darden, *The Earth in the Looking Glass* (New York: Anchor Press, 1974).

19. See, for example: Wallace E. Pratt, *Oil in the Earth* (Lawrence: University of Kansas Press, 1942). Kirtley F. Mather, *Enough and to Spare* (New York: Harper, 1944). Kirtley F. Mather, "Petroleum Today and Tomorrow," *Science* 106 (1947): 603–9; and *The Advancement of Science* 4 (1948): 292–300.

Knowledge
about Nature

To LIVE IN THE NEW WORLD with all its trouble and challenge requires much knowledge of many different kinds. There is consequently deep interest in the process of knowing, and its various aspects have frequently been analyzed and classified. Quite commonly there is recognition of two significantly different ways whereby knowledge may be gained. These are variously defined or designated, depending largely upon the mental attitude and particular purpose of the person who is making the analysis and constructing the classification. The distinction between subjective knowledge and objective knowledge, for example, is by no means the same as that between conceptual knowledge and perceptual knowledge. Both of these binary groupings of the many diverse factors in the complex process of knowing are useful, each in its own way. They focus attention, however, upon mental processes and tend to neglect the important distinction between the kinds of knowledge that may be gained, whatever the procedures used in the search.

To facilitate our present inquiry concerning the nature of the universe, it seems desirable to consider the various kinds of knowledge in two major categories which may

be designated as scientific knowledge and spiritual knowledge. The former may be described simply as knowledge
about those aspects of the universe, its matter and energy
and their transformations, which can be measured in relation to space and time; it is the kind of knowledge sought
by the scientist *qua* scientist, whether he be a physical,
biological, or social scientist. Spiritual knowledge may be
described as knowledge of those aspects of the universe,
and of life within it, which are inherently nonmeasurable
in relation to space and time; it deals with spiritual factors,
intangible but none the less real: the sense of beauty caught
by the artist, the awe and reverence of the religionist, the
ideals of righteousness and the loyalty of the man of integrity, the essence of personality in individuals, and the
characteristics of the universal creative power.

These two kinds of knowledge are intimately interrelated. Indeed, there seems to be no hard and fast water-
tight boundary separating them. The categories constructed
by human minds often do violence to the continuity of
nature. Only by a more detailed scrutiny of each can the
distinction between them be appraised. Scientific knowledge will therefore be examined in this chapter, spiritual
knowledge in the next.

Many different methods are used in the various scientific
disciplines to facilitate the search for more information and
better understanding. The procedures that have proved
successful in the application of new knowledge to increase
human efficiency and comfort are often different from those
used in fundamental research. Indeed, the ability to design
an effective program for specific purposes in research and
development is an essential qualification of a successful
scientist or technologist. Even so, there is a basic methodology that undergirds all aspects of the search for scientific knowledge and technology. It is an intellectual tool,
fashioned and perfected by man in comparatively recent
years. It has both objective and subjective aspects; it de-

pends upon perceptual abilities of the human mind and senses, but it necessarily involves conceptual thinking. Abundant experience in using it has led to the recognition of four significant operations that ordinarily are performed in a well-established sequence.

The first of these is *observation*. Convinced that the senses can report accurately to the mind the nature of the external world, the observer makes a conscious effort to note all relevant facts about the material objects under study. He tries to identify significant characteristics of form, composition, movement, and transformation under various conditions. He watches the behavior of animals and plants in diverse circumstances and takes note of the changes in position of inanimate objects. Seldom is the operation a casual one. To be useful in the quest for scientific knowledge, the observation must be more than a passing look at something. It must be made analytically, with intent to see things as they actually are, to gain insight concerning their real nature. Not only must the senses be trained to discriminate between minutely different characteristics, but interest in the particular objects of study must be quickened to keep the mind alert and attentive. Good observers are in great demand but in short supply, even among highly literate populations.

The second of the four steps in the methodology of science involves controlled *experimentation*. From the innumerable and completely interrelated varieties of nature, one is selected, or a very few variables are isolated so that causal relationships may be discovered. As before, the inquiry is specifically directed toward the identification of facts or the recognition of events.

The search for factual data has necessitated increasing refinements of methods of observation, with ever greater precision of measurement. Characteristically, there has been an increasing use of mechanical, optical, and electronic apparatus for making and recording observations. This is not only to gain precision of measurement beyond that obtainable with the unaided eye but also to minimize the human factor in this phase of the entire operation. No matter how impersonal, objective, neutral, and unemotional a scientist of integrity may strive to be, the elements of fatigue, if not of personal predilection, are always lurking in the shadows. The nervous system of the most cold-blooded, unprejudiced, and honest human being is less reliable than the well-designated automatic recording and computing instruments available today.

In significant contrast, the third step in scientific research involves human capabilities for which there can never be any mechanical substitute. This is the *construction of logical theories* to explain in terms of cause and effect, or of other spatial or temporal relationships, the factual data which have been compiled. Therein is displayed one of the unique characteristics of the human being, one of the glorious capacities of mankind. It matters little whether the logical theory is a concept as all-inclusive and abstruse as Einstein's general theory of relativity or as specific and practical as the garage mechanic's notion that a short circuit in defective wiring is the cause of failure of the starting mechanism in your car. The construction of any concept by postulation is indicative of the value and significance of intellectual activities in cosmic history.

Deductively formulated theories must seem logical to those who conceive them, else they are unworthy of consideration. But the logic of such a theory is not the test of its validity. Hence it is essential in the quest for knowledge to take the fourth step, *validation,* in the sequence which is now firmly established as characteristic of what we call the scientific habit of mind. The logic theory must be tested by the accuracy of its predictions. If the theory is true, then under prescribed conditions things will happen in an expected way. Hence there must be additional precise observations and further controlled experimentation. There-

ticles possessing the inferred properties. Thereafter the pace of nuclear research quickened. Each forward step in the advance of knowledge not only provided a new base from which to move onward but also put new tools in the hands of the researcher. Chadwick's "stream of neutrons" is a prime example.

Late in 1938, Otto Hahn[8] and Fritz Strassman,[9] using such a stream in their laboratory in Germany, made the epochal discovery that the nuclei of uranium atoms actually broke apart, with release of some of their binding energy, when impinged upon by neutrons moving at suitable velocities. The concept of nuclear fission, for some time only a plausible speculation, was thus demonstrated to be valid. Almost immediately thereafter, Lisa Meitner[10] and her nephew, Otto Frisch,[11] established the fact that there could be a self-sustaining chain reaction involving nuclear fission on such a scale as to provide a source of energy greater than any previously known. To carry the knowledge about this new possibility beyond the reach of Nazi tyranny, Miss Meitner hurried away from the laboratory where she had been working with Dr. Hahn to continue her research with Niels Bohr[12] in Denmark. As a matter of fact, the principle of the chain reaction, building up the energy liberated by nuclear fission, was independently developed by several groups of experimenters, including Leo Szillard[13] and W. H. Zinn[14] in the United States, during the first months of 1939.

Thus far the experiments had been on a microscopic scale. Indeed, the faith of the nuclear physicists that energy could be released from uranium nuclei in usable amounts was still based on only partial evidence. Not until December 2, 1942, was that faith justified by works. Very few persons knew about the epoch-making event at the time; all such things were under wartime security regulations. But the records were kept in meticulous detail, the memories of those present were vividly persistent, and the

fore, also, the scientist whose imaginative mind developed the theory must announce to his fellow-scientists not only the concept he has postulated but also the nature of the factual data on which it was based and his method of testing its validity. Others then can make the observations and perform the experiments which will be for them the test of its truthfulness.

I have referred to the ideas originating in the minds of men when they take the third step in the process of gaining scientific knowledge as logical theories. Actually, the idea may at first be only a *speculation*—sometimes wild or naive—or an intuitive hunch. With further thought, it may be discarded as unworthy of serious consideration, or retained, with or without modification, to become an *hypothesis*, suitable for thorough testing in the fourth step (validation) of the scientific method. Only after at least partial testing and validation does an hypothesis achieve the status of a *theory*.

In common practice, the hypotheses used by any scientist to explain any particular set of observations or the results of any specific experiments are commonly ideas that he has acquired from his teachers, his colleagues, or his reading of relevant literature. Occasionally he may construct a new hypothesis by analogy with one previously applied under different circumstances; more rarely he may actually originate a truly new idea. In any event, experience has taught us that it is well to give consideration to many possible hypotheses, rather than to only the first one that comes to mind, in the search for a logical theory. The method of multiple working hypotheses[1] has been widely used in the scientific enterprise since the last decade of the nineteenth century.

After a *theory* has been extensively tested by many scientists and its principle repeatedly validated under a variety of conditions, it may attain the status of a scientific *doctrine*; especially if it can be formulated as a mathematical equation, it may also be known as a *law of nature*. For example, the idea that the force of gravity attracting any two bodies varies directly as the sum of their masses and inversely as the square of the distance between them, first put forward as a somewhat tentative hypothesis[2] by Isaac Newton in 1666, is now universally recognized as one of nature's laws. Newton's *Laws of Motion*, a keystone of classical physics, have not been repealed by modern physics; relativity theory and the new knowledge about sub-atomic entities have merely defined more sharply the limits within which those laws are sovereign. Similarly, the idea that all animals and plants have come into existence through the orderly processes of organic evolution, published somewhat reluctantly by Charles Darwin in 1859, is today one of the most widely recognized doctrines of the life sciences.

The next noun in this sequential designation of concepts—*speculation, hypothesis, theory, doctrine*—would be *dogma*, a doctrine proclaimed by an establishment—ecclesiastical, academic, political, or whatever—to be immune from question or reexamination. But dogma is, or should be, anathema to all scientists and technologists. Such was the commitment made by the Royal Society of London— founded in 1660 and today the oldest and probably the most prestigious of all scientific organizations—when in 1663 it adopted as its motto, "Nullius in Verba." That motto derives from a couplet penned by the Roman poet, Horace, which may be translated to read "And do not ask me, by chance, what leader I follow or what godhead guards me. I am not bound to revere the word of any particular master."[3]

This mental attitude has been maintained, with few exceptions, by all who use the methods of modern science in the search for knowledge about nature and man. Darwin's theory of organic evolution, for example, was significantly emended during the hundred years following his first announcement of it, especially as a result of the expansion in knowledge about genetics which came after his death. Today's doctrine of organic evolution is sometimes labelled "Neo-Darwinian," or better "a modern synthesis."[4] It remains wide open to critical analysis and further emendation by any competent person. It has been validated by so many facts of observation and experimentation that it is now something more than a theory; it is a doctrine, but never a dogma.

All of which means that the kind of knowledge gained by scientific research is in the public domain. It is available to anyone who equips himself with the appropriate intellectual and instrumental tools of research. The results of its application may be marvelous but they are not miraculous in the sense that they transcend all comprehension. It is moreover especially significant that this knowledge is always dynamic, never static. Perhaps the most praiseworthy feature of a scientific theory is not that it explains many things or that it provides great economy of thought by integrating large numbers of facts within a general principle; it is rather that it inevitably stimulates further inquiry, that it opens new vistas for still more adventures along the path toward greater knowledge and truer understanding.

It is furthermore implicit in this procedure that the final arbiter is the set of facts made known by observation and experiment. In supporting or demolishing a scientific theory, the scientist cites the facts themselves as evidence rather than the opinion of a respected leader in his field or the pronouncement of a famous scholar of bygone days. Many scholars contemporary with Galileo refused to accept the conclusions drawn from his experiments with falling bodies because these conclusions contraverted the laws of nature enunciated by Aristotle. Their attitude is considered today as utterly naive, to say the least. The modern scientist seeks to discover the laws of nature as they actually are, not as he or anyone else thinks they ought to be.

This attitude toward knowledge has permeated far beyond the confines of white tiled laboratories and ivy covered walls. In a technological culture such as ours, everyone has use for mechanical devices of greater or lesser complexity. Machines are obdurate, neutral, impersonal. They breed respect for facts, the facts of physics, chemistry, and engineering, if not indeed the facts of life. Thanks to science and technology, the universe is no longer something to be feared, but instead something to be understood.

None of us is completely devoid of wishful thinking in our moments of contemplation. There is still much unquestioning acceptance of the pronouncements of others, especially of those in official or professional positions which enable them to influence public opinion. Nevertheless, the appeal to facts for guidance and judgment rather than to the dictum of an ancient sage or the dogmatic assertion of a man who wields great economic or political power is a fairly common practice nowadays.

The history of the scientific enterprise indicates unmistakably that the quest for knowledge about nature has been greatly facilitated by freedom of communication between its devotees, regardless of political, ethnic, or creedal barriers that might tend to isolate them from each other. Since the dawn of what we call civilization, no nation has had a monopoly of ideas. There were times in the past when new understandings were gained, new concepts formulated, by scholars working quite independently of their fellows in foreign lands. Aristotle (384–322 B.C.) and Archimedes (287–212 B.C.), for example, were completely unaware of the scholarly activities of Confucius (550–478 B.C.) and the generations of Chinese savants who followed him. Since the Middle Ages, however, most of the significant advances in knowledge have been made by persons who were well acquainted with what was going on in their fields of competence and interest throughout the entire world. Unquestionably, the free flow of ideas is the essential life-blood of science. The long series of studies, sketched briefly in Chapter 1, which eventually made nuclear energy available for human use, is an especially pertinent case in point. The first atomic bomb might appropriately have been stamped "Made in America," but ideas essential to its manufacture had come from France, Germany, Italy, Denmark, and Great Britain. Interdependence in science is so well recognized that most scientists the world around are firmly convinced of the value of international scientific organizations and believe strongly in the sharing of knowledge and theories without regard for political boundary lines.

The methodology of science has proved to be the most effective tool in gaining knowledge about the world around us. It may also be applied to the study of men as individuals or as members of a society. The four steps outlined above are still applicable and must continue to be followed meticulously. In many such investigations, however, controlled experimentation is necessarily minimized or outlawed altogether by common consent. Statistical studies take its place to a considerable extent. Thus the demographer can predict with a high degree of accuracy how many persons of a certain age in a particular population will be alive ten years from now. He makes no pretense of knowing in advance which individuals in the group will have died.

Actually, nearly or quite all of the so-called laws of nature are now known to be statistical laws. The principles of thermodynamics are good examples: they set forth with great probability the statistical consequences of the more or less random behavior of vast numbers of individual molecules. Thus, you can increase the air pressure of your automobile tire by forcing into it additional molecules of atmospheric gas so that the number of impacts of the individual molecules against the tire's confining walls per unit of time is increased. Or you can get the same result by heating the air in the tire and thus increasing the average velocity of its molecules so that, again, there are more impacts per unit of time. There is always an increase in temperature of the tire and its contents when a car that has been standing for a time is driven at ordinary speeds over ordinary highways, because of friction between tire and road and within the tire itself as it flexes to give you a smooth ride. Consequently, a weak tire is more likely to blow out when an automobile is moving swiftly than when it is standing still. The principles of thermodynamics have significant application for us all. It matters little whether a few of those molecules of air greatly exceed the average velocity of impact of the lot; their idiosyncrasies will be offset by others which fall far below that average. An exception or two, here or there, does not violate the orderly results of the thermodynamic principle as it applies to the entire mass.

In contrast, when dealing with human beings as the unit objects of scientific investigation, the exceptional individuals cause trouble. Regulations pertaining to human nature are much less amenable to scientific research than regulations relating to machines. Actually, each human being is unique as an object of scientific study, except of course for identical twins reared in the same environment. Without taking on at this point the thorny problem of free will, it suffices to note that a human being is an organized system, far more complicated than the molecules of gas to which the statistical laws of thermodynamics apply. Moreover, the statistics concerning man pertain only to thousands or millions of individuals in an aggregate, whereas those of thermodynamics pertain to billions or trillions or even more. Consequently, departures from the average have greater significance. Even so, the search for knowledge about man has already been rewarded by the discovery of some very important generalizations which seem adequate to connect large numbers of particular facts. There is good reason to expect that with further research the science of man will continue to grow in comprehension. The limits of knowledge have not yet been reached in this or any other field in which the methodology of science can be applied.

There are ultimate limits, however, for this kind of knowledge. One of its most significant limitations is implicit in the description I have given of the first two steps in the process of scientific research. To be useful in the all-important third step, each observation and each controlled experiment must involve measurements, the more precise the better. In the last analysis, science can deal only with objects, forces, and motions which can be measured. In other words, it deals only with space-time relations of material entities and the forces that change those relations. The knowledge it gains is quantitative knowledge. Fundamentally, it is knowledge *about* the nature of the universe and of man.

But the universe and man have qualitative as well as quantitative aspects. Insistently, men crave understanding of the meaning of human life. They ask not only what and where and how but also *why*. Note how the nuclear physicists sometimes point with pride to a new theory and call it elegant. Classical physicists denoted their theories as concise, or all-embracing, or logical, or simply adequate. The adjective *elegant* implies something qualitative. It suggests that the theory is a clue to the real meaning of events, not merely a description of those events in terms of a general principle.

In a deep philosophic sense the more we learn about nature, the more mysterious nature seems to be. It is no longer possible to portray adequately the structure of an atom, as Niels Bohr did in the 1930s, by constructing a mechanical model with marbles and wire. No longer can a well-educated man conceive light as simply waves in motion or streaming corpuscles. Nor does it appear that the mysteries of quantum electrodynamics, nuclear spin, and the nature of light are mysteries only for those inadequately trained in mathematics.

There is growing skepticism among scientists concerning the ability of the scientific method to deal effectively with the ultimate questions bound to arise in the quest for a satisfactory philosophy by which one's daily life may be directed. Henry Margenau, Professor of Physics and Natural Philosophy in Yale University, expresses this skepticism when he writes: "Since knowing is only part of human experience, science is limited. For our total experience includes, besides knowledge, such components as feeling, judging, willing, and acting as well."[5] Albert Einstein put a similar thought in his characteristic style:

> The scientific method can teach us nothing else beyond how facts are related to, and conditioned by each other. . . . The knowledge of truth as such is wonderful, but it is so little capable of acting as a guide that it cannot prove even the justification and the value of the aspiration toward that very knowledge of truth. Here we face, therefore, the limits of the purely rational conception of our existence.[6]

Occupying the pulpit in his hometown church on a Layman's Sunday, Warren Weaver, 1954 President of the American Association for the Advancement of Science, included this statement in his remarks:

> Science is a highly successful way of handling certain limited aspects of man's experience—chiefly the behavior of inanimate and animate nature insofar as this behavior can be usefully measured and described with numbers. But science—modern science as recognized by its best practitioners—makes no pretense of possessing the rules or procedures for dealing with many

other aspects of man's experience, namely the more significant matters which deal with our recognition of values, of beauty, of duty, and of ultimate purpose.[7]

That the kind of knowledge gained by scientific research is inadequate to meet the total needs of deeply troubled man is also indicated by some very practical considerations. The hope that expansion of this kind of knowledge will eventually resolve all human ills has been fractured if not completely shattered by the grim facts of human behavior. Increased efficiency resulting from the application of the fruits of scientific research has not been generally accompanied by any comparable improvement in morality. By and large, there seems to be no correlation between efficiency and righteousness.

Not long after the close of World War II, I visited the notorious concentration camp at Dachau, near Munich in Germany. By that time the extermination center had been pretty well cleaned up and was being made into a sort of shrine. I stepped inside one of the gas chambers to inspect its mechanical equipment. Then I entered the crematorium with its brick furnaces against which rested withered wreaths of flowers tied with faded ribbons, placed there by those who had reason to believe that their loved ones had been exterminated in that place of death. Outside were small wooden buildings like overgrown doghouses where the ashes from the furnaces had been stored until trucks could take them to the chemical works. It was a marvel of scientific efficiency: automatically sealed doors, gas ducts of superb design, furnaces that displayed ultra-modern principles of thermodynamics, conversion of all valuable products. But just beyond was a large bank of earth cloaked with yellow flowers and a sign, "Eighteen thousand unidentifiable bodies are in this mass grave." Even in the bright sunlight of that lovely August afternoon, I could almost literally feel the weight of the black cloud of man's inhumanity to man, increased to the nth degree by the efficiency of science. No! At Dachau certainly there was no correlation between scientific efficiency and righteousness.

The truth is, of course, that the techniques and devices made available by science and technology may be used either for good or evil, for human betterment or degradation. In the arena of morals and ethics they are neutral. Knowledge about the transformation of matter into energy, for example, may be used either to produce the uncontrollable, lightning-swift chain reaction of nuclear fission of plutonium or uranium in a weapon of incredible destructiveness, or to provide man with a vast source of power for peaceful purposes by controlled chain reactions of exactly the same fundamental nature. Obviously, the uranium atom is completely amoral; it is what men do with it that has moral and ethical significance.

The neutrality of scientific knowledge and of the instruments it makes available for use has long been recognized by workers in the physical sciences and accepted for many aspects of the biological sciences. It also applies to the social sciences, although there are many who believe that expansion of human knowledge in this field of research will by itself be adequate eventually to save mankind from evil ways. It is doubtless true that the development of the social sciences has thus far been prompted in the main by the desire to promote human welfare. Most social scientists are inspired by fine motives. Nevertheless, it is all too evident that the new knowledge of human behavior may be used to serve bad as well as good ends. Scientific methods of propaganda have been applied with great success by autocratic dictators to pervert large numbers of citizens in certain countries. Dachau could never have been what it was, had not someone used with diabolical skill the modern techniques whereby propagandists win friends and make fools of people.[8] Modern advertising campaigns are using the recently organized knowledge of social psychology for ends which may be detrimental to social progress, as well as for highly commendable purposes. The use of the new social techniques by governments and pressure groups may prove to be more effective in the evolution of society in these times of rapid change than economic structures or social stratification. Even here, the knowledge and tools of science reveal themselves only as means to an end. And the end may be either beneficent or malevolent, depending upon the purpose toward which the technicians direct their efforts.

Undergirding the quest for scientific knowledge, whether in the physical, the biological, or the social sciences, there is a comprehensive fundamental doctrine or principle to which all men of science give allegiance. It is the established conviction that transformations of matter and energy within the space-time framework of the universe always and everywhere take place in an orderly manner in accordance with regulations that are, at least potentially, comprehensible by human minds. If that doctrine be valid, the universe of which we are a part is a universe of law and order, a cosmos, not a chaos. Many of the laws are statistical, dealing with large aggregations of individual entities, such as the regulations controlling the behavior of gases to which reference has already been made. Many, and in the last analysis, perhaps all, are laws of probability, such as those that are practically useful in the study and application of genetics. Some may be stated with mathematical precision, such as Newton's Second Law, $f = ma$ (force is equal to a mass times its acceleration), or Einstein's Universal Law of Equivalence, $E = mc^2$, referred to on page 11. On occasion, enclaves of disorder may appear within the ken of observational science, but more meticulous study and deeper analysis have often discovered that disorder was only apparent and not really a violation of the principle of law and order. Here, significantly, as soon as subatomic particles, with their apparently uncertain activities (the Heisenberg principle), are organized to form atoms, the atoms are subject to regulations that are being spelled out with increasing precision in many a physico-chemical laboratory. No matter how chaotic or turbulent a local situation may temporarily seem to be, whether in a nuclear research establishment or in a conflict of self-interests in human affairs,there is no question in the minds of scientists but that we are living in a law-abiding universe.

Knowledge about the laws and regulations of nature does not, however, give man mastery over nature. Rather, it enables him to humor nature, to adapt his activities so that they harmonize more perfectly with those laws and regulations. Using that knowledge, he may do more efficiently what he wants to do, but only within the limitations established by them. Knowledge is power, but the possession of power does not of itself guarantee that it will be wisely used. It may well be doubted whether the most complete knowledge about nature would alone make man competent to solve all the problems of our present time of trouble.

NOTES

1. Thomas C. Chamberlin, "The Method of Multiple Working Hypotheses," *Journal of Geology* 5(1897): 837–48; reprinted 39(1939): 155–65.

2. cf. J. B. Cohen, "Newton, Isaac," *Dictionary of Scientific Biography* (New York: Scribners, 1974) 10: 42–101.

3. E. N. da C. Andrade, *A Brief History of the Royal Society* (London: The Royal Society, 1960).

4. Julian Huxley, *Evolution: The Modern Synthesis* (London: G. Allen & Unwin, Ltd.)

5. Henry Margenau, *Open Vistas* (New Haven: Yale University Press, 1961), 3.

6. Albert Einstein, *Out of My Later Years* (New York: Philosophical Library, 1950). Also to be found in *Ideas and Opinions* (New York: Crown Publishers, 1952) 41 and 42.

7. Warren Weaver, An address delivered in the Congregational Church in New Milford, Ct., May 16, 1954. Privately circulated.

8. cf. Alfred M. Lee, *How to Understand Propaganda* (Rinehart, 1952).

3

Knowledge
of Nature

THE QUEST FOR THE KIND of knowledge to which the designation spiritual knowledge may be given is at least as old as the search for scientific knowledge. It is implicit in many of the legends and myths of antiquity. It has led, over and again, to the sense of kinship with nature which is so commonly found in primitive cultures. The myths of ancient times are replete with references to Mother Earth. The idea that man was somehow an offspring of the earth is found in almost every primitive religion.

Take for example the Greek legend involving the wrestling match between Hercules, the epitome of physical prowess and indomitable courage, and Antaeus, son of Gea, the earth goddess. It is a legend dear to the heart of the geologist inasmuch as the name of his science comes from that designation for the earth plus the Greek word *logos*, one of the meanings of which is study. Thus geology is the study of the earth.

In that ancient story, the two antagonists met in a hand-to-hand battle to the death. Each time the mighty Hercules gained the upper hand and was about to throttle his opponent as he lay almost breathless on the ground, Antaeus drew renewed strength from his Mother Earth and sprang up to continue the conflict. At last, Hercules held Antaeus high above his head, breaking Antaeus's contact with the earth, and there delivered the coup-de-grace. Although in most of his other feats of strength and valor Hercules was a hero, in this particular episode he was the villain. Antaeus was mankind personified; Hercules personified all the forces and conditions antagonistic to human welfare.

In that legend there is a lesson for modern men. As long as we keep our feet on the ground, maintain our sense of kinship with the earth, and have in mind our dependence upon our planet for our sustenance and strength, we can meet undaunted all "the slings and arrows of outrageous fortune." Let that spirit of relationship be forgotten or denied, we too may suffer the gory fate of Antaeus. Is not this in reality the thought that motivates the contemporary conservationists who plead for the preservation of wilderness areas, even more than the desire to provide places for research concerning species of fauna and flora that otherwise might become extinct?

Down through the ages of human history there seems always to have been at least a dim apprehension of aspects of reality that are not amenable to physical measurements. The uniqueness of man among the creatures of the earth is indicated by such profoundly and manifestly true statements as the oft-quoted "Man does not live by bread alone." There is something in human nature which craves expression in works of art, discovers and establishes moral standards, demands obedience to ethical principles. Expressed in the vocabulary of the cultural anthropologist, man "is an evaluating animal, always and everywhere. Moral standards of some kind always exist."[1]

Apprehension of esthetic, moral, and ethical standards, principles, or values is spiritual knowledge as contrasted with scientific knowledge. It involves the qualitative factors in human nature and in the universe as distinct from the quantitative. The descriptive terms used in discussing this kind of knowledge are quality words, like lovely or ugly, inspiring or debasing, noble or sordid, righteous or evil, rather than quantity words, like millimeter or mile, gram or ton, microsecond or day. Its concepts can never be expressed in terms of metrical equations. Yet its relevance to human behavior is inescapable. Men may not always live in accordance with its finest principles, but their lives are always influenced to some extent by its concepts.

It is appropriate to call this kind of knowledge spiritual knowledge. If matter may be defined as that something no two parts of which can occupy the same space at the same time, and I know no better definition, then spirit may be defined as that something which cannot be described in terms of space and time. Thus, spiritual knowledge differs from scientific knowledge most significantly because it deals with wholly different data.

It is helpful also to note that spiritual knowledge is essentially knowledge *of* something, rather than knowledge *about* something. A professional connoisseur of art may be told many things about a particular painting that he is asked to appraise, whether it is on canvas or on paper, whether oils or crayons were used, whether it is of the classical or the surrealist or whatever other school, even whether it was the work of a world-famous painter or of an unknown tyro. But he will refuse to pass judgment on its beauty or its artistic merit until he has actually seen and studied the picture itself. He must know whether it has a meaningful message that awakens something within himself, whether it causes or fails to cause a spirit of admiration to stir his inner being. That is to say, he must gain knowledge of the painting rather than merely have knowledge about it before venturing an appraisal.

A gripping poem about the beauty of the sunset and the glory of the dawn may alert the reader or the listener to the possibilities of a future experience, but it is not until he pauses at sundown or sunrise on some favorable vantage point in quiet, open-minded contemplation that he becomes aware of the deeper meaning of those words or finds himself responding to the wonders of nature in ways that transcend all description. The old-time Christian evangelist displayed real wisdom when he exhorted his listeners to "come *to* Jesus" after he told them many things *about* Jesus.

This kind of knowledge, therefore, is personal or private knowledge. In this respect also it differs from scientific knowledge which by its very nature is in the public domain. Molecules react in the same way under a given set of circumstances, regardless of political loyalties, ethnic origins, or geographical location of the experimenting scientists. A complete description of the experiment and its results is all another scientist needs to know in order to evaluate a colleague's work. Each technician does his best to keep his own predilections, emotions, and mind-set completely apart from his observations and experiments. Only in the construction of deductively formulated theories do these subjective factors enter into the process of developing scientific knowledge, and even there the scientist tries to be as objective as he can.

In the quest for spiritual knowledge, however, a large part of the data, perhaps even all of the data are to be found in the reactions of persons to aspects of nature, including human nature, of which they become aware. Individuals react in various ways to quite similar circumstances. They respond differently to stimuli that seem to be practically identical. Subjective factors are of prime importance. Still more significant is the fact that no person can convey by descriptive words to any other person the full meaning of an experience through which he gained spiritual insight. Each one must depend to a great extent upon his own experiences if he is to grow in spiritual knowledge.

This is of course both the glory and the danger of the process. For many it has been such a stumbling block that they have denied the validity of all subjective knowledge and relegated all so-called spiritual realities to the limbo of untrustworthy myth or neurotic hallucination. Quite certainly, human beings have often had experiences that are meaningless in reference to spiritual knowledge and have as often interpreted their experiences in terms which can only be designated as rubbish. The wise seeker after spiritual knowledge therefore borrows a leaf from the notebook of his scientific comrades in zealous research and adopts the policy that only those experiences are useful in his quest "which can be ordered in terms of a system of spiritual values," to use the cogent expression of James B. Conant.[2]

The key words in that expression are *order* and *system.* They remind us that the universe within which we are a part is a cosmos characterized by harmonious regulations which dictate rational processes, not a chaos characterized by haphazard, discordant sequences of unrelated events. It may be assumed confidently that there is spiritual law in the natural world. Those words also suggest that there may well be a close parallel between the process of gaining spiritual knowledge and the process of developing scientific knowledge.

A scientist encountering a problem in his research, or puzzled by a series of factual observations for which he is trying to construct some unifying concept, may have his moment of insight. Archimedes, so the story goes, leaped from his bath and ran naked through the streets of Athens shouting "Eureka!" because he had discovered the law of floating bodies. Many a modern man of science has reported moments of insight during which the solution of a long-perplexing problem has come to him suddenly as a flash of inspiration out of the blue. Similarly, religious prophets, worried by the iniquities of their clientele, or pondering the alternatives of possible activities which they might recommend, have had their moments of insight. Moses returns from a brief sojourn on a lonely mountain with the ten commandments which he reports were dictated to him by Jehovah. Many a Judeo-Christian prophet of biblical and later time has reported that in a moment of deep contemplation or in a dream he heard the voice of an angel of the Lord, or even of God himself, giving him some good advice or the solution to a problem. The man of science reports the result of his moment of insight as a discovery, the man of religion as a revelation. Their moments of insight are probably very much alike with regard to the mental processes involved before and during the flash of inspiration, despite the different vocabularies they use.

To take the first step in the process of gaining spiritual knowledge, introspection is substituted for observation. Oriental philosophers have developed techniques of introspection that are only recently becoming commonly known in the western world. In the meeting of East and West, the West has much to learn as well as to teach. Yet even the first naive attempts of a western man to comprehend his own nature and discover the deeper meaning of his own life are likely to yield results of precious import. There seems to be a "sensitivity to qualities felt through an inner responsiveness, through intuitive apprehension that goes directly to reality."[3] Introspection never fails to uncover the essentially human characteristics of allegiance to some type of "values that cannot be demonstrated scientifically or weighed or measured, but well up out of inner experience."[4]

The process of intuitive apprehension is of special concern to research scientists. Here I am not thinking primarily of the distinction appropriately made between concepts by intuition and concepts by postulation. According to F. S. C. Northrop, "a concept by intuition is one the complete meaning of which is given by some thing directly inspected or immediately apprehended."[5] This is involved in every report conveyed to us by our senses concerning the apparent nature of things in the world around us. For example, we sense immediately and directly that grass is green. "A concept by postulation is one the meaning of which in whole or part is proposed for it syntactically, formally and theoretically by the postulates of some explicit, deductively formulated theory."[5] The concept of wave-lengths in Maxwell's electromagnetic theory is a good example.

It is in a different sense from this that I am thinking of intuitive apprehension in the context of the process of gaining spiritual knowledge. Many a scientist reports that a concept by postulation has come to him as though it were a sudden flash of insight. At times even when his conscious mind is not directly focused on the problem that has long troubled him, the illuminating idea abruptly intrudes itself. With a great feeling of joy, he suddenly grasps the answer which had hitherto eluded him. Archimedes in his bath is the classical illustration I have already given. Something similar is acknowledged by almost every introspective seeker of scientific knowledge about whom such intimate details of the intellectual life have been communicated to others.[6] Evidently, intuition should not be casually rejected as something characteristic only of feminine idiosyncrasies.

Moving onward to the second step in the process of gaining spiritual knowledge: instead of controlled experiments, the student makes himself acquainted with the spiritual experience of others. Here is a wealth of data. Biographies, especially autobiographies, of many men and women who have experienced something believed to be significantly revealing about the human spirit are available from numerous times and places. Bible stories are a valuable part, but only a part, of the impressive library. Especially fruitful also are intimate contacts with persons of commendable behavior and character. In such associations there is often communication on a deeper level than that of words and gesture.

Pursuing this second phase of the process, one promptly encounters an embarrassment of riches. Many interpretations of experiences are recounted as though they were the facts themselves. Contradictory interpretations are at least as numerous as those which harmonize with each other. How can one intelligently screen out the trash and the error from the valuable and the true? Or more basically, in these days of psychoanalysis, how can one have confidence at all in any spiritual interpretations of human experience, even in one's own?

Underneath the active consciousness of every one of us there indubitably are hidden desires, hatreds, or frustrations which influence powerfully all we think as well as all we do. Many religious concepts, the Freudian psychologists tell us, are simply projections of human longings and fears into the vast system of nature, the persistence of a youthful father-image, the outgrowth of the very natural human desire for protection and support. Cultural anthropologists have already demonstrated that the concepts which meet with approval, whether of a spiritual or a scientific nature, are profoundly influenced by the local and temporary culture. Attitudes and beliefs, as well as activities, respond to, and to some extent result from, the conditioning which plays so large a role in human education from the cradle to the grave.

Edmund Sinnott has dealt effectively with this problem in his admirable book, *Two Roads to Truth.*[7] He points to the fact that "the concept of rationalization is a two-edged sword. . . . What makes a Freudian an agnostic is as certainly dependent on physiology and as resultant from conditioning as what makes someone else a man of faith. . . . Rationalizing, like conditioning and physiological factors,

may be called on to explain skepticism as well as faith."
With Gordon Allport we must agree that "origins can tell
nothing about the validity of a belief."[8]

Actually, I have already given the clue to the way in
which the results of introspection and meditation, whether
one's own inquiry or those of other people, are to be appraised. Only those experiences are useful which have
meaning in terms of an orderly system of spiritual values.
This principle carries us forward to the third step in the
process of gaining spiritual knowledge. To apprehend the
spiritual aspects of life, it is necessary to seek general laws
connecting a number of facts of human experience, even
as it is essential in scientific research to construct logical
theories which explain relationships among a number of
particular facts of human observation. It is in terms of such
general laws that the great prophetic voices of religion
have made their proclamations. Note, however, that moral
and ethical postulates, by their very nature, cannot be
firmly established on only one particular fact or experience. How large the number of such facts must be, no one
can designate in advance, partly because we are dealing
with qualities rather than quantities. The parallel is close,
moreover, to scientific postulates which also by their nature cannot be produced inductively for a single fact or
from a limited set of facts.

The parallel continues. The third step must be followed
by a fourth, regardless of what kind of knowledge we seek.
Before a general principle pertaining to the spiritual aspects of life can be accepted as trustworthy, it must be
appraised in terms of the consequences of its application,
both in the personal and in the social life of man. Pragmatic
tests, similar to those habitually applied to the principles
and hypotheses discovered by scientists, must also be applied to the principles and precepts revealed to moralists
and theologians. This is not easy. It is, for example, not
yet clear whether the enlightened self-interest of a well-

meaning humanist is inferior to, or equal to, the unselfish commitment of a devout theist as a stimulus and guide toward spiritual growth and social integration. Perhaps in the long run, it will be found that each has glimpsed a different facet of the ineffable truth. In any event, it is by the fruits of their spiritual knowledge, rather than by the roots, that they will be known.

A man kills his wife and four small children, and, when charged with murder, claims that he is innocent because he had heard the voice of God telling him to kill them. Even those best acquainted with him are quick to say that his alleged spiritual experience was phony and that he is a psychopathic person. The pragmatic test is customarily applied in such horrifying cases; it should also be applied in the routine appraisal of all postulated spiritual knowledge.

In these recent years of re-examination, an unusually large number of young people—and many older persons as well—have been withdrawing from the long-established religious organizations to set up splinter sects and to embark upon new crusades of one kind or another. The trend is found not only among Protestant denominations, Jewish synagogues, and Roman Catholic churches, but also among the Greek and Russian Orthodox assemblages and the adherents of the Muslim, Hindu, and Buddhist faiths. Characteristically, each of the new sects proclaims, either implicitly or vociferously, that it is the possessor and guardian of truth. Competition among them has often led to friction, especially when one group seeks to increase its numbers by proselyting among others. How should the precepts, doctrines, and dogmas of such sects be evaluated? Certainly not by quoting carefully selected excerpts from the Bible, the Talmud, the Koran, the teachings of Gautama Buddha, or the sayings of Confucius; nor by the degree of charisma displayed by key persons in the group, commonly designated by such titles as Leader, Teacher,

Guru, Pastor, Sister, Brother. Appraisal of the worth and truthfulness—or lack thereof—of the tenets promulgated by any religious sect is a vital part of the fourth step in the process of gaining spiritual knowledge. That appraisal is made by observing the day-by-day activities, the customary behavior, the life-style of the members of the group. Is the human behavior, motivated by the doctrines and dogmas of the sect, conducive to the long-continuing welfare of mankind? Or, do those beliefs provide an easy excuse for escape from the world of reality?

It is in this fourth step, rather than the third, that the well-known and much-discussed categorical imperative of Immanuel Kant has its valid application, as Henry Margenau has pointed out.[9] Kant's maxim, "Act so that the motive of your action can be willed to be a universal law," is misconstrued, says Margenau, if it is regarded as a moral code; it is in fact a rule for validation.

> This is at once apparent when we watch the categorical imperative in action. Suppose a person is tempted to lie. If he overcomes the temptation, he has obviously satisfied the imperative, at least in the sense of its author who used it to show that lying is immoral. But if the person tempted does lie, he can justify his course without much trouble by appealing to the uniqueness of his situation—for every situation is unique—and by averring that . . . it really ought to be a law for people to lie in this particular situation. Thus the categorical imperative neither sanctions nor interdicts lying. Its real significance rests in its suitability as a method for verifying an already chosen norm. It really says "Suppose you lie; then see if society can exist if everybody lies." This is a matter for empirical test.[9]

The postulational and confirming phases of ethics must

be clearly distinguished one from the other. The kind of analysis customary in the sciences emphasizes the significance of the fourth step in the process of gaining spiritual knowledge.

Such analysis furthermore drives home the necessity for more of this kind of knowledge and more widespread recognition of its practical effects. The activities of human beings are largely governed by ideas and ideals. Ideas are created in human minds. Ideals are established by cosmic regulations and conditions. The ideals discovered by man, largely through experiences that are essentially religious, are at least as influential as the ideas stemming from scientific observation.

In recent years many scientists have expressed the thought that the qualitative aspects of human life should be investigated by scientific methods even as are the quantitative aspects. The extension of knowledge in psychology and psychiatry, combined with the progress toward understanding in the social sciences, encourages them in this belief. Thus it is said that ethics should be developed scientifically and values should be determined on a scientific basis. There is much truth in this idea.

I believe, however, that a distinction should be made between the scientific *attitude* and the scientific *method* in this phase of our quest for knowledge. Quite certainly the scientific attitude should be adopted in all inquiries concerning spiritual knowledge. But, as I have tried to indicate, the methods used in this part of the search for truth are inherently different from those that have proved so successful in extending scientific knowledge.

In the scientific attitude toward any problem the mind is always open to novel ideas. Outrageous hypotheses have occasionally proven true in many a scientific break-through. Truth in science has often been astonishing to the scientists concerned. It may be so even in the realm of the spirit. Any doctrine, no matter how venerable or widely accepted, is subject to new appraisal and possible revision. Always the final court of appeal is found in the observations and experiences of men and women. Adopting the scientific attitude in seeking truth concerning the qualitative aspects of human life is essential, but this does not mean that the methods to be used are identical with the ones applicable to the quantitative aspects.

To meet the challenge of these swiftly changing years, both kinds of knowledge are needed. As so often proves to be the case, the answer here is not either/or. Wisdom demands both scientific knowledge and spiritual knowledge. Integration of the two is not only desirable but possible. Indeed, it is essential if we are to grow in wisdom.

Such integration is expedited when we consider the process of gaining spiritual knowledge as a dynamic, self-correcting discipline. In religion, as in science, it is possible to keep oneself open to new insights and at the same time to hold fervent convictions. The adventures of the spirit pursue a different route from that traversed in the adventures of the mind. Tension between faith and reason is a catalytic agent which stimulates the strengthening of character. To realize the full potentialities of a human life, a person must seek truth and pursue it wherever it may be found. Not mind alone, but spirit also, is essential to the fine art of noble living. Together they make a person whole.

NOTES

1. Clyde K. Kluckholm, *Proceedings of the Stillwater Conference on the Nature of Concepts* (New York: Foundation for Integrated Education, 1950), 86.

2. James B. Conant, *Modern Science and Modern Man* (New York: Columbia University Press, 1952), 99.

3. Edmund W. Sinnott, *Two Roads to Truth* (New York: Viking Press, 1953), 212.

4. Edmund W. Sinnott, *Two Roads to Truth*, 208.

5. F. S. C. Northrop, *Proceedings of the Stillwater Conference on*

the Nature of Concepts (New York: Foundation for Integrated Education, 1950), 38.

6. R. M. MacIver, ed., *The Hour of Insight* (New York: Harper Brother, 1954).

7. Edmund W. Sinnott, *Two Roads to Truth*, 200–203.

8. Gordon Allport, *The Individual and His Religion* (New York: Macmillan Paperbacks Edition, 1960), 109.

9. Henry Margenau, *Proceedings of the Stillwater Conference on the Nature of Concepts* (New York: Foundation for Integrated Education, 1950), 112–13.

$$\boxed{4}$$

The Grand Strategy
of Evolution

THE DOCTRINE OF EVOLUTION, with its vast breadth
and unfathomed depth, is comparatively new in the
thoughts of man. It antedates Darwin, but only within the
century following the publication of his *Origin of Species*
in 1859 was its relevance to all phases of cosmic history
as well as to the development of life upon earth generally
recognized. In any attempt to understand the nature of
the universe and of man, knowledge will provide significant clues to the characteristics, both qualitative and quantitative, of the power or powers operating by means of
that process.

Surely if the spiritual aspects of the universe are to be
found anywhere, it is in the activities and development of
the living creatures known to us here on the earth. The
life sciences are numerous and many of them have become
highly socialized in recent years. Each of them offers opportunity to gain insights that bear upon both the process
and the power involved in evolution. The devotee of each
scientific discipline, whether it be psychology or paleontology, genetics or endocrinology, will naturally approach
the problem from his own particular viewpoint. Here,
however, the geologist who concerns himself with the his-

tory of life as well as with the history of mountains and plains, oceans and continents, metalliferous ores and mineral fuels, has a special responsibility.

He may stand, so to speak, on a reviewing platform and watch the procession of living creatures as it emerges from the darkness before the dawn of the Cambrian Period, more than half a billion years ago, and moves forward along the corridors of time into the bright light of this noontide of science. The record of geologic life development is far from complete. To become a fossil is actually a relatively rare achievement for animals and plants. For a fossil to be found and to attain the distinction of a place in a museum display or study collection is an even less likely circumstance. Consequently, many of the marchers in that procession remain invisible and their characteristics must be inferred from those of their predecessors and descendants. Nevertheless, enough is now known to permit the observer on that reviewing stand to report at least the broad outlines of the drama unfolding before him.

The oldest known rocks contain no fossils whatever. Their age may now be determined by recently available methods involving the radioactive timekeepers. These indicate that many of the rocks now exposed at the earth's surface were formed at least two billion years ago and that some of them are nearly four billion years old.

The oldest known fossils are in rocks which date back to approximately two billion years ago. They consist almost exclusively of the remains of the simplest forms of plant life; most of them are the characteristic secretions of calcareous algae and the carbonized tissues of seaweeds. Animals are indicated only by the presence of small structures in certain sandstones which look like the worm tubes seen on modern beaches where worms have bored their way into the wet sand. What kind of worm-like creatures they may have been cannot now even be surmised.

These indications that life was present in the very ancient seas are the basis for the names geologists give to the first two of the five great eras[1] into which for convenience they divide the recorded part of earth history: Archeozoic Era, the era of archaic life, and Proterozoic Era, the era of primitive life. (See Geologist's Timetable, Appendix 1.) Certainly the life in the seas of those eras was far different from, and much more lowly than, the life of later time. So far as we now know, the most highly developed inhabitants of the land areas of those ancient times were primitive plants reproducing by means of spores, presumably ancestors of the mosses or ferns.

Approximately six hundred million years ago the procession of the living passed the milestone which marks the beginning of the Cambrian Period, the first period of the Paleozoic Era. In rocks formed at that time there is an abundant and amazing array of fossils. Cambrian strata record the presence of representatives of every important phylum of invertebrate animals and of spore-bearing plants. The animals ranged from one-celled protozoans to richly endowed trilobites. Brachiopods were probably the most numerous and cephalopods the most powerful inhabitants of the Cambrian seas, but trilobites were the most intelligent. Like other arthropods they possessed segmented bodies, paired appendages, eyes, and antennae. Their nervous systems were adequate to control and coordinate the activities of many highly specialized organs. In general, however, the Cambrian animals of each major kind were inferior creatures when compared with their successors in later periods. Especially notable also is the complete absence of vertebrate animals and flowering plants, so abundant today.

Moving swiftly through the successive periods of time within the Paleozoic Era, we note in passing such events as the advent of the first fishlike animals in the Ordovician Period and of the first air-breathing land dwellers—snail-like creatures—in the Silurian Period. During the latter period and in the subsequent Devonian Period, the fishes deployed into many orders, such as sharks, rays, garpikes, and lungfish. Some of the fish with a well-developed air bladder may have used that organ, originally intended primarily to control buoyancy, as an accessory breathing organ or lung, even as do the modern lungfish.

Before the end of Devonian time, the first amphibians made their appearance. In the transition from aquatic, gill-breathing fish to terrestrial, lung-breathing quadrupeds, fins evolved into legs, but gills did not become lungs. It was from the bladder into which air was swallowed that lungs developed. Only a few quite generalized amphibians are known to have lived in the next period, the Mississippian, but many kinds, some with fairly long and slender legs, are recorded in the Pennsylvanian Period. All of them, like frogs and salamanders of today, must have laid their eggs in water. When first hatched, their young presumably breathed by means of gills, but as they matured they went through the tadpole-frog metamorphosis to become air-breathing, land-dwelling quadrupeds. Late in the Pennsylvanian Period or at the very start of Permian time, the first reptiles made their appearance. Reptiles lay their eggs on the ground, and the lungs of their young are fully functioning when hatched. The skeletons of some of the early Permian reptiles are scarcely distinguishable from those of some of the amphibians. The evolution of reptiles from amphibians can hardly be doubted.

The Permian Period is the last subdivision of the Paleozoic Era. By the close of that era, about 225 million years ago, every major kind of animal life except birds and mammals had made its appearance. To be sure, the insects which were numerous in the later Paleozoic periods were only roaches, dragonflies, and similar types. Bees, ants, and butterflies had not yet come into being, but the insect class had established itself in the mundane scheme of things. Similarly, crustaceans of the modern type such as crabs and lobsters were still wanting from the array of arthropods, but the trilobites and other primitive crustaceans had firmly established the role of animals with a carapace. Indeed, the trilobites seem somehow to have overdone the matter. They had flourished in the Cambrian Period, deployed rapidly to a climax of many varieties in the Ordovician, then had gone into a slow decline to become entirely extinct before the end of the Permian Period.

The Plant Kingdom, too, has left a noteworthy record of Paleozoic development. Certain Upper Cambrian shales contain the spores of mosses or primitive ferns. There were tree-ferns in Devonian time, and the Pennsylvanian coal measures contain the record of a lush flora which must have clothed many a land mass with an evergreen mantle. Many of the trees and shrubs of late Paleozoic time were true seed-bearers, mostly of the cycad and gymnosperm types. Many were tree-ferns and lepidophytes. But among them were some which were completely transitional in nature, a type of seed-fern representing the link in the chain of evolution from reproduction by means of spores to reproduction by means of seeds, marking one of the most important advances in the development of plants. These intermediate forms disappeared by the end of Permian time, but the record of their temporary presence remains to the present day.

Passing on to the Mesozoic Era, it is necessary to do little more than recall the fact that the approximately 155 million years encompassed by it are popularly referred to as The Age of Reptiles. Dinosaurs first appeared in the Triassic Period, deployed into herbivorous and carnivorous types in the Jurassic Period, and maintained their mastery of the land throughout the long Cretaceous Period. Among them were some of the largest, most viciously powerful, and most heavily armored brutes that ever stalked across the face of the earth. Reptiles also became rulers of the sea and lords of the air in the latter half of the era, when plesiosaurs and ichthyosaurs were the scourges of the deep and predaceous pterosaurs soared freely in the sky.

Among the widely variant reptiles of the Permian Period were two kinds which displayed skeletal characteristics foreshadowing those of mammals. One of these lived in what is now the southwestern interior of the United States, the other in South Africa. Only in Africa did this development continue. Fossil bones in Triassic rocks of that continent indicate the transition from mammal-like reptiles to primitive, presumably egg-laying, mammals. Jurassic rocks in Eurasia and North America as well as in Africa contain fossil jawbones and teeth closely resembling those of certain modern oviparous and marsupial mammals. And in the lower Cretaceous rocks of Asia have been found the fossil skulls of at least two kinds of placental mammals. They were relatively small quadrupeds, more or less similar to the tree-shrews of today.

Birds also made their appearance in the Jurassic Period. They are not the descendants of the flying reptiles, the pterosaurs, but of some early type of small, light-limbed, swiftly running reptile closely akin to the first of the dinosaurs. Indeed, had not the imprint of feathers accompanied the bones of one of the first specimens found, it would undoubtedly have been classified as a reptile rather than as a bird. Once more there is an actual record of a form transitional between two of the apparently widely separated classes of creatures living today. Most of the few known Mesozoic birds had teeth and lacked the skeletal support for powerful wing muscles which characterizes all modern flying birds.

Had any of us been able to look at the earth from some distant vantage point in late Cretaceous time, seventy-five or eighty million years ago, we would probably have turned away in sorrow and disgust. "Surely life on earth has gone mad in its lust for power gained by strength of talon and claw," we would say. "Cold-blooded, small-brained, huge-bodied dinosaurs run rampant over the land; equally rapacious reptiles lord it over their fellow creatures at sea and in the air. How can any good come from such a world as this?"

But have patience. There soon came a change, as all things, cosmic and terrestrial, change. The milestone on the path of time marking the end of the Mesozoic Era and the beginning of the Cenozoic Era commemorates the complete extinction of the entire horde of dominant reptilian brutes. Dinosaurs vanished from the land; icthyosaurs and plesiosaurs disappeared from the seas; pterosaurs made their exit from the skies. Modern reptiles are but the remnant of a vanquished army.

Early in the Cenozoic Era, placental mammals spread rapidly from their Asiatic birthplace over all the continents except Australia which has had no land connection with any other continent since the Cretaceous Period. Equipped as they were with four-chambered hearts and brains well nourished by warm blood, they promptly diverged into many specialized types: hoofed herbivores, clawed carnivores, rodents, sloths, primates, and so on. Their dominance during the succeeding periods of the Cenozoic Era is the basis for the frequent reference to its seventy million years as The Age of Mammals.

Many of the experiments in which reptiles were involved during the Mesozoic Era seem to have been repeated by the mammals during Cenozoic time. The creatures were different and environments were not the same, but the local and temporary objectives were closely similar. Some of the earlier herbivorous mammals, for example, seem to have sought security for themselves in terms of huge bulk and efficient defensive armament. The Uintatheres and Amplypods of Eocene time were just as truly victims of megalomania as such sauropod dinosaurs as Brontosaurus and Diplodocus had been, long before in mid-Mesozoic time. Those huge mammalian quadrupeds, far larger than any modern elephant, persisted for fifteen or twenty million years, developed massive bony "horns," and then disappeared from the lands of the earth before the middle of the Cenozoic Era. Glyptodons developed a massive bony carapace covering their entire bodies in much the same fashion as the box in which turtles now gain protection from their enemies. They flourished for a time in South America and spread into North America in later epochs of the Cenozoic Era, but they became extinct during the Great Ice Age, a few tens of thousands of years ago.

By and large, the most successful of the mammals seem to have been those which combined a high degree of intelligence with great agility and flexibility of movement. Ability to approach stealthily and leap suddenly to overpower their prey with strong claws and sharp teeth had great survival value for the carnivorous quadrupeds, and these characteristics reached their climax in such felines as the sabre-toothed tiger. Ability to spring quickly into swift flight saved the life of many an herbivorous mammal. Competition for speed must have had much to do with the evolution of the single-toed horse from its five-toed ancestor. Skeletal structures and muscular strength would, however, be of little avail were it not for the brain and nervous system that control the movements of the body. Awareness of danger, or of prey, must be keen and must be followed promptly by appropriate action. Intelligence, as well as physique, became in the mammals an especially important factor in the screening operations of natural selection.

It is noteworthy, too, that the great majority of the mammals that lived through the trials and vicissitudes of the Cenozoic Era and gave rise to the present mammalian population of the earth were creatures skilled in the fine art of cooperation. The wolf pack is more likely to be successful in providing food for its members than is the lone wolf seeking sustenance for itself in a solitary chase. Cattle lived together in herds long before men gathered them into corrals or barnyards. Ability to perfect a social organization seems to have had great survival value at many different levels of development and most conspicuously so among the widely variant mammals.

But in such comments I am pushing beyond the bounds of mere description into the enticing area of interpretation of the fossil record. It is one thing to describe the successive inhabitants of the earth in ages past. All competent students must agree concerning the observable record. It is quite a different thing to interpret that record in an attempt to discover its vast meaning. Here are likely to be sharp differences of opinion; general agreement is hardly to be found. What follows therefore should be taken as one man's opinion, the interpretation and meaning which seem to me to be valid but which may be accepted or rejected by others.

Most people will agree that the record demonstrates the progressive nature of the all-inclusive process of life development. Not only have many kinds of organisms developed increased complexity of organic structure with the passage of time, but those organs and structures have meant greater efficiency in obtaining food, provided increased protection against enemies, and have made possible more adequate care of immature offspring. More fundamental still, the record indicates an expanding awareness of conditions and forces in the environment. Antennae extend slightly the radius of effectiveness of the sense of touch; eyes permit a creature to recognize objects and movements at relatively great distances. Concurrently, there has been progressively an increase in the number of ways whereby a creature may react to external stimuli or display by its activities its own essential nature. Compare, for example, the number of possible reactions to external

conditions that might be displayed by a Cambrian brachiopod, a Devonian lungfish, a Permian reptile, a Jurassic
bird, or an early Cenozoic mammal. You will note that I
am measuring progress in mathematical terms with no
reference to whether living creatures have become more
noble or more lovely or have attained higher morality.
Applying a mathematical yardstick to the products of the
evolutionary process—and no other means of evaluation—we find that the evolution of life during geologic
time has been marked by truly progressive achievement.

Surveying the record of that achievement, we naturally
seek general principles that governed the changes it encompasses. It is easy to select fossil bones or shells from
successive time sequences of rocks and display them in
museums or books in such a way as to give the impression
that they represent lineages developing in straight lines
from a primitive ancestor to a more modern type. Such
carefully selected arrangements strongly suggest the concept of orthogenesis—evolution proceeding directly and
unerringly in precisely determined directions.

The evolution of horses during the Cenozoic Era is a
well-known example. It has given rise to an often quoted
ditty:

> Little Eohippus was no bigger than a fox,
> And on three toes he scampered over Tertiary rocks.
> "But," said little Eohippus, "I am going to be a horse,
> And on my middle fingernails I'll run my earthly
> course."

It is quite true that Eohippus, the Dawn Horse, was about
the same size as a fox, which it rather closely resembled
in general form. It is also true that it walked almost on the
flat sole of its foot with three small hooves on its rear feet
and four on its forward ones. Moreover, in successive epochs of Tertiary time there appeared progressively larger

horses with longer legs and necks and with a change in bone structure making possible a posture that might be described as standing or running on the tip of the middle digit. The lateral digits gradually decreased in size and finally dwindled almost to nothing, so that the modern horse has only a single hoof representing the tip of the middle digit on each foot. But Eohippus could not possibly have had a vision in his mind's eye of his splendid equine descendant of late Tertiary and Recent time. He had enough to do to maintain his own existence in the midst of the perils of his own day; he could give no thought to a morrow, ten million or more years away.

The concept of orthogenesis has arisen in the minds of men as an imposition on nature; it is not a valid principle or law of evolution. If any geometric pattern can actually be found in the record of life development, it is one of radial evolution rather than of linear progression. There is nothing to indicate that lineages are impelled by some internal or supernal force to keep on evolving indefinitely in the same direction.[2]

The progressive development of terrestrial life through geologic time has been accomplished by a long series of minuscule steps. Each enduring change in anatomy or behavior appears to improve the adjustment to the total complex of environmental factors at the time and place in order that a particular kind of life may have a better chance to survive. Such changes seem to be a result of the application of the method of trial and error—many failures and an occasional success. The procedure is that of experimentation; it often resembles the purposeful activity of the well-qualified scientist in certain phases of his laboratory research and pilot development. Whatever may be the shortcomings of the available raw material—essentially the characteristics of the gene pools involved—these are used with apparent cleverness to accomplish each forward step of the procession along the path of life. The transition from paired-fin, gill-breathing, aquatic vertebrates with incipient air bladders to quadrupedal, lung-breathing, terrestrial vertebrates is one of the countless examples. To characterize such methods as opportunistic is not in any sense to demean them. "You will know them by their fruits."

Any purpose implicit in the evolution of a new species from its antecedent species is short-range. One step may lead to another, but no one of them suggests any prevision of the specific demands and adjustments that will occur in the distant future. The geological records afford no suggestion, for example, that a blueprint for man had been drafted on any architect's drawing board a half-billion years ago. Even as late as fifty million years ago, in early Tertiary time, there is no hint of any design to produce a creature precisely in the anatomical mold of man as he has emerged in glacial and post-glacial time.

The endeavor of each species of animal or plant to maintain the existence of its own kind of life resulted in progressive improvements in the adjustments of anatomy and behavior to the environment in which it lived. Changes in environment imposed new conditions to which further adjustments had to be made if the species was to survive. Changes thus effected frequently add up to such an extent that the biologist or paleontologist must designate the modified creature as belonging to a new and different species. Thousands of times as many species as are living today must have lived in the geologic past, only to become extinct.

But extinction is not of itself a mark of failure. Speaking figuratively, every species of animal or plant has made its appearance on the stage of life; there it has played its role in the ineffable drama. For a moment it may be center stage, but sooner or later it must make its exit; having had its day, it ceases to be. There are however two exits from the stage of life. Through one of these the dinosaurs, for example, went out into oblivion. No drop of dinosaur blood, or speaking more accurately, no gene from the dinosaur gene pool can be found in any living animal. This, I suppose, might be called a failure. In contrast, Eohippus left the stage of life through the other exit. In doing so, he handed on his torch to offspring who were more like modern horses than he was. There must be many genes inherited from Eohippus in the chromosomes of modern horses. Surely no one could say that Eohippus, although extinct, was a failure.

The universal urge for every type of life to maintain the continuity of its own existence, come what may, has often been accompanied by a spirit of adventure, or has given rise to that spirit. Many times when a particular type of life has made itself fairly secure in a congenial environment, it has done its utmost to expand into other environments. As soon as the reptiles had consolidated their position as lords of the land, they sent expeditionary forces into the sea where aquatic saurians became the most powerful creatures of the deep in Jurassic and Cretaceous times. Like Alexander longing for more worlds to conquer, the reptiles also invaded the air. The pterosaurs were far more numerous in the Cretaceous Period than birds.

Such an interpretation of the geologic record has obvious teleological implications. Teleology is a philosophical concept conveying ideas of purposive activity, of an explicit endeavor to attain previously established goals, usually on a long-range basis. In the sciences dealing with inanimate objects there is certainly no place for such ideas, and teleology is now a naughty word for many scientists. Students of the life sciences, however, should not thoughtlessly discard it out of hand. When a squirrel collects acorns and stores them away, in order that it may have a food supply during the snowy winter, its behavior certainly has teleological implications. To say that the squirrel behaves that way because of its instincts is begging the question. Why did the first squirrel for the first time do something to provide for its survival in a future emergency? Whenever a behavioral scientist uses the words "in order that" rather than "because of," his explanation of an event has teleological implications. In the processes of evolution the goal-seeking is on a short-range basis, but it is certainly there.

Flying reptiles evolved from one type of terrestrial reptiles, birds from another quite different type, and bats from terrestrial mammals. In each instance the changes in body form resulted in a new ability—that of long sustained flight in the lower levels of the atmosphere. Although the achievement thus gained was the same, the means used were different. The wings of pterosaurs, birds, and bats are utterly different in structure. Within the regulations governing the processes of evolution, such as the general directive to "be fruitful and multiply and replenish the earth" (including land and sea and sky), great flexibility in responses is permissible.

Even so, certain patterns of form and behavior recur in endless repetition. When terrestrial reptiles were modified to live successfully in the water, their bodies were streamlined and their legs became flippers. Some of them even imitated their remote piscine ancestors by developing a median fin. Much later the land mammals also essayed the conquest of the oceans. Although dolphins are descendants of land mammals, they have a remarkable, albeit superficial, resemblance to ichthyosaurs, which were the offspring of land reptiles. In this effort to adjust to a particular environment by developing a similar body form, there is no record of an actual reversal of the process of evolution. Breathing by means of gills would surely have been a great convenience for aquatic reptiles, but that habit and structure had been abandoned long before by their amphibian ancestors. All aquatic reptiles and mammals must come to the surface of the water to breathe air. Similarly, the eggs of reptiles must be laid on land, although the eggs of their ancestors, the amphibians and fishes, were deposited in water. A structure once lost is never regained. All aquatic reptiles must either come out of the water and lay their eggs on land or bring forth their young alive. Functions, however, are very different from structures in this regard. Legs became flippers; they did not revert to fins. But flippers perform the same function as fins, even though their structure is quite different.

Thus far we have been considering things that fall into the category of scientific knowledge. The question naturally arises concerning the possibility that the grand strategy of evolution embraces also the development of aspects of reality which interest specifically the seeker after spiritual knowledge. Most of such items pertain to the nature and activities of human beings, and they will be considered in the next chapter. We have learned, however, that the beginnings of many a structure or process which emerges in full flower in modern times may be traced far back in time and far down on the scale of life. Spiritual aspects of life can leave no fossil trace, but they play an important role in the interpersonal relations of men, brought together in groups or organized in societies. It is possible, therefore, to interpret certain records of geologic life development in terms that have spiritual significance.

The first animals and plants were individualists, some of them sufficiently rugged to keep alive amid the vicissitudes of Precambrian time. From the unconscious, inorganic compounds of the earth's surficial zone where mineral matter, water, and air intermingle had emerged self-sustaining cells capable of growth and reproduction. Drawing upon the available resources of matter and energy, these primordial creatures developed organic structures and modes of living specifically designed to maintain existence for themselves and their kind. By the close of the Cambrian period, some of them—the corals, for example—were living side by side in colonies, although there is nothing to suggest that in a coral colony there is any social organization involving mutual aid among its constituent polyps.

No one can say with certainty when animals first began to care for their own offspring. Possibly the trilobite parent protected and guided her immature young until they were well able to fend for themselves. Mesozoic reptiles almost certainly rushed to the assistance of mates and offspring in time of danger even as do some of the modern reptiles. It was not, however, until birds and mammals had appeared in late Mesozoic and Cenozoic time that anything approaching mother love, as we know it today, emerged as a life-sustaining extension of self-consciousness.

Progress from self-consciousness to social-consciousness is most conspicuously displayed among invertebrates by such insects as ants and bees and among vertebrates by birds and mammals. This relatively recent result of evolutionary processes seems to have been independently developed in each of these three widely separated kinds of creatures. It is therefore indicative of a trend inherent in those processes. Insect societies reached their present high level of complex organization and efficiency some ten million years ago, in late Tertiary time. Human societies have been in the process of organization only within the last few hundred thousand years.

Willingness to sacrifice one's self for one's offspring is generally rated as highly commendable. To sacrifice one's self for the benefit of some unrelated individual is looked upon with only a little less universal approval. Mutual aid and self-sacrifice are products of evolution. Both emerged as living creatures moved forward along the corridor of geologic time.[3] Judged in terms of the values which men customarily hold dear, the conclusion seems clear that evolution has resulted in progress toward the attainment of "the good, the true, and the beautiful." No strictly mathematical evaluation of evolutionary development can encompass its totality.

This poses a real problem. It is doubtful whether the principles of heredity as discovered by geneticists include any directive toward progress. The mechanisms of inheritance, as we know them today, guarantee change but not improvement. Offspring will differ from their parents but there is no certainty that they will be superior or inferior. In the grand strategy of evolution, natural selection must be responsible not only for the development of new species but also for the progressive nature of the whole operation. Those creatures survive and reproduce their kind which have hereditary equipment permitting and causing them to make the most effective life-sustaining adjustments to the sum total of environmental factors. Over the years, those adjustments have involved not only the gradual emergence of new organs, structures, and functions of individual bodies, but also the increasing reliance on mutual aid and the progressive attainment of societal organization.

Here, too, the stimulus of changes in environment must be especially important. The history of each continent has been characterized by a succession of expanding and retreating seas which have flooded more or less of its surface. Relatively small embayments of the epicontinental seas, like Hudson Bay today, were commonly present within the borders of the continents at times when the land areas were most extensive. Each embayment provided a more or less unique environment for the shallow-water marine fauna that inhabited it, thus offering opportunity for experimentation with new organic structures and novel modes of existence. These were tested under local conditions, and some proved to have great survival value. When the ocean waters expanded toward continental interiors, the hitherto separated embayments coalesced to spread vast mediterranean seas from ocean basin to ocean basin. The local champions were pitted against each other in the continental sweepstakes. Those members of the several provincial faunas best equipped to survive in the new environment constituted the cosmopolitan fauna of the widespread intracontinental seaways. Later, the seas retreated, and again there were only narrow shelf-seas and marginal embayments on the edges of the continents. Once more there was a stimulus for local champions to try new ways in each environmental province.

There have been at least a score of these irregular cycles of sea advance and retreat since the beginning of the Paleozoic Era. Alternation between provincialism and cosmopolitanism seems to have been a significant factor in evolution.[4] It has of course been effective in the evolution of the inhabitants of the land as well as of the shallow seas. When sea-ways crisscrossed the continents, the lands were separated into many provinces. When seas retreated, terrestrial creatures entered a cosmopolitan phase in their life history. The evolution of land mammals cannot be understood without taking into consideration the presence and absence at various times of land bridges between the two Americas and between North America and Eurasia.

Especially for the nonmarine creatures, changes in climate have frequently been a compelling stimulus for evolutionary progress. For example, the evolution of the first air-breathing quadrupeds from their gill-breathing aquatic ancestors apparently occurred under the compulsion of seasonal dryness in late Devonian time. When rivers perennially dwindled to a succession of stagnant pools, the survival value of an accessory air bladder capable of serving as a respiratory organ, and paired fins suitable for locomotion on the ground, must have had great significance. Adversity has often been the stimulus for innovation in the evolution of life.

Once the handicaps or limitations of ancestral lineage

are overcome and the old organs effectively modified for new functions, opportunities for abundant experimentation have always been at hand. The development of amphibians and reptiles from those first venturesome quadrupeds of the Devonian Period has left an amazing record of widely differentiated kinds of terrestrial vertebrates in the late Paleozoic rocks. Or observe the history of the mammals. The lungs of their reptilian ancestors had replaced gills functionally but not structurally. The air bladder from which lungs developed had been so related to the blood circulation system that the head and therefore the brain was not nearly so well nourished as the rest of the body. That handicap was removed when the primitive mammals appeared in Mesozoic time with their four-chambered hearts. The survival value of intelligence made possible by well-nourished brains was especially important in subsequent periods. The deployment of land mammals during Tertiary time is one of the most spectacular episodes in the whole record of ancient life development.

Anyone who looks at this record in broad perspective, seeking to find something systematic in the all-encompassing process of organic evolution, can hardly fail to gain insights of far-reaching philosophic significance. By and large, the activities responsible for the observed results seem to have been closely akin to the methods of an expert scientist. The method of trial and error, with occasional successes, is revealed again and again. Versatility and ingenuity characterize the operations at every stage. Apparently unpromising materials have yielded commendable products. The process of evolution commands awe as well as admiration.

NOTES

1. George C. Simpson, "Some Problems of Vertebrate Paleontology," *Science* 133 (1961): 1679–89.

2. P. Kropotkin, *Mutual Aid, A Factor in Evolution* (New York: Knopf, 1914); W. C. Allee, *Cooperation Among Animals* (New York: Schuman, 1951).

3. Kirtley F. Mather, "Geologic Factors in Organic Evolution," *Ohio Journal of Science* 24 (1924): 117–45.

5

The Evolution
of Mankind

MAN IS A MAMMAL, equipped with a four-chambered heart, mammary glands, and a well-nourished brain. More specifically, he is a placental mammal, giving birth to young after a considerable period of gestation within the body of the female parent. His ancestral lineage can be traced far backward in time. His morphologic and physiologic evolution has been in accordance with the same laws or principles as those to which all other animals are subject; and it will continue so to be.

Within the amazingly diversified group of placental mammals, man belongs to the order of Primates, in which he is grouped with tarsiers, lemurs, monkeys, and apes. All possess nails at the tips of fingers and toes rather than claws or hooves, and have brains that are large in relation to the dimensions of their bodies. Most of them still have the primitive number (five) of digits in hands and feet. Any differences in their biological development are differences in degree rather than in kind.

The first of the Primates to appear in the geologic record lived in early Tertiary time, about sixty million years ago. Their fossilized bones and teeth have been found at many places in North and South America, Eurasia, and Africa.

Most of the species that have been recognized and named are referred to the genus *Notharctus*. They closely resembled the tarsiers or lemurs now living in Madagascar and eastern Africa as far as anatomical forms and structures are concerned. Presumably their habits of life were also similar. Certainly their teeth indicate that they could have been omnivorous in their food habits, even as are most living nonhuman Primates, specializing neither as flesh eaters nor vegetarians, but preferring eggs, fruits, and insects.

Among the notharctines were the ancestors of all living Primates, many of whom are notably different from their progenitors in various significant ways. Thus there appeared in mid-Tertiary time, twenty to thirty million years ago, the first known representatives of the sub-order Anthropoidea, the taxonomic group embracing the anthropoid apes—gibbon, chimpanzee, orangutan and gorilla—and mankind. Fossil bones and teeth of these earliest anthropoids have been found at several widely scattered localities in Eurasia and Africa but in no other continent. Among them is the especially interesting fossil found in the middle 1960s in the Fayum desert in Egypt. It was in rocks dated as twenty-seven million years old and is named *Aegyptopithecus*. According to E. L. Simons,[1] it "almost certainly lies in or near the ancestry of *Dryopithecus*, the modern Great Apes and man." Several species of the extinct genus *Dryopithecus* are known from Miocene strata in Egypt and India. They are so generalized in body structure and tooth pattern that they apparently include the progenitors both of man and of the great apes of Africa and Borneo. They are definitely not in any line of evolutionary development that could have led to modern monkeys.

Among the most informative of the dryopithecine fossils are the specimens found in the Siwalik Hills of India and West Pakistan. Some of these seem to be in the ancestral lineage of the Hominidae rather than that of the Pongidae (the anthropoid apes).

In the classification scheme of modern taxonomists, Hominidae is the scientific designation for the family of mankind, including extinct as well as living genera and species. The earliest Hominids in the geologic record appear between fifteen and five million years ago. The wide spread of that dating is due not so much to lack of precision in determining the age of the relevant fossils as to uncertainty concerning the biological relationships of the creatures they represent. In the last fifty years, hundreds of specimens of extinct anthropoids have been found in Kenya and Tanzania,[2] in southern Asia and in eastern China, in geologic formations ranging in age from thirty million to a half million years ago. Some of them can be referred by the experts without hesitation to the extinct group of dryopithecines. Others are referred with equal confidence to an extinct group of hominoids known as the australopithecines, of which the genus *Australopithecus* is the chief representative. These are so specialized in body structure and tooth pattern that they must have been in the ancestral lineage of modern man and could not have been ancestral to any existing ape. Still others, however, have such different characteristics, or are so fragmentary and incomplete, that the experts are divided into two camps; some would regard them as advanced members of the family Pongidae (apes), others as primitive hominoids. The transitional nature of these specimens (many of which were found in India and Kenya and are now referred to the extinct genus *Ramapithecus*) reinforces the belief, widely held by anthropologists and paleontologists, that man evolved from ape-like ancestors in East Africa and South Asia during the past ten million years of earth history. Thence stone-age man spread to Europe, other parts of Africa and Asia, and eventually to North and South America and Australia.

In that historical development, the members of the genus *Australopithecus* deserve special attention. They lived during the Pliocene Epoch (extending from about thirteen million years ago to one or two million years ago), the last epoch of the Tertiary Period. They are best known from fossils found in eastern and southern Africa. The name Australopithecus means Southern Ape, but they almost certainly belong in the ancestral lineage of mankind, not in that of the modern anthropoid apes. They stood and walked erect; their facial features had a distinctly simian cast, but their brain capacity was intermediate between that of modern man and that of chimpanzees and gorillas. Two of the species of this genus have recently been described by John T. Robinson[3] on the basis of his long and detailed study of fossils found in South Africa. One is called *Homo africanus* by Robinson but is referred to as *Australopithecus africanus* by the majority of physical anthropologists. It is believed by many of the experts to have been directly ancestral to modern man. The other, called *Paranthropus robustus* by Robinson, but *Australopithecus robustus* by many others, was more ape-like and is believed to have become extinct by mid-Pleistocene time.

This evolutionary divergence between the ancestral lineage leading to modern man and that leading to the chimpanzee, gorilla, and orangutan is marked by the adjustment of such creatures as *A. africanus* to a new way of life on broad savannas, tropical or subtropical grasslands with only scattered trees and shrubs, whereas the members of the other lineage clung to the old ways of arboreal or semi-arboreal life in forests and jungles. In their new way of life, their social organization was even more essential to their survival than it was for their simian relatives in the much safer arboreal environment. Naked, unarmed and alone, a member of man's ancestral lineage must have been a rather helpless creature, no match for a carnivorous feline, scarcely able to secure adequate food for himself. The australopithecines must have organized themselves in small, tightly knit troops, at least as well structured as are the troops of chimpanzees investigated by modern students of animal behavior. Presumably they used clubs and stones for attack and defense; some of them may even have been tool-makers as well as tool-users, although no flaked stone tools or weapons or other artifacts are known with certainty to be associated with their fossil bones.[4]

The belief that australopithecines were the ancestors of mankind has recently been further strengthened by the work of two of my Harvard colleagues, Bryan Patterson and Arnold D. Lewis, pursuing field investigations and laboratory studies under the auspices of the Museum of Comparative Zoology.[5] Since 1963 they have been concentrating their attention on Miocene and Pliocene fossil mammals of northern Kenya. Exploring for fossils in Pliocene strata on the flanks of Lothogam Hill in 1967, Lewis found a portion of the right half of a lower jaw (with a tooth intact) of a creature subsequently identified by Patterson as an early member of the human family, a hominid that lived among the australopithecines approximately five million years ago.

Of even greater significance are the fossils found in the early 1970s by Richard Leakey (son of the late L. S. B. Leakey), east of Lake Rudolf in northern Kenya, and by Don Johansen and his associates at Hadar in the Afar Triangle of Ethiopia. In each of these regions, the paleoanthropologists found fossil bones of sub-human hominoids that date back to three million years or more ago. Among Johanson's finds are the remains of a female creature, about twenty years in age, consisting not only of skull and jaw fragments but of nearly forty percent of the entire skeleton, including parts of the pelvis, vertebral column, ribs, arms, and legs. Nicknamed Lucy by her discoverers, she stood about three and a half feet tall, walked erect, was definitely a sub-human hominid rather than an australopithecine, and lived three or four million years ago.[6] It is probable that after further study the experts will decide to place

Lucy not only in a new species but in a new genus as well. Her precise place in the family tree of mankind may not be known, however, until additional fossil hominids are found in the rich hunting grounds for man's ancestors in Kenya and Ethiopia. Such discoveries will almost certainly be made within the next few years.

In the meantime, we can now say with confidence that representatives of genus Homo first appear in the known geologic record either near the close of the Tertiary Period or very early in the Pleistocene Epoch, The Great Ice Age, of the Quaternary Period. Paleontologists and anthropologists have recently revised the nomenclature of prehistoric hominids, discarding many of the names previously used for isolated fragmentary fossils and reflecting the modern consensus concerning the affinities of well-known fossil creatures to each other and to modern man. Thus, *Pithecanthropus erectus* (the famous Ape-man of Java, named by Du Bois in 1892) and several other creatures known or believed to have had closely similar characteristics are now known as *Homo erectus,* an extinct species in the genus to which modern man belongs, rather than a species in an extinct genus, closely related to but separate from the one that includes us. Similarly, *Sinanthropus pekinensis* (the Man of China, found in 1929 near Peking) is now known as *Homo pekinensis,* with the same implication of closer affinities to other species in the genus than had earlier been inferred. Much the same changes in presumed relationship have overtaken the older nomenclature for extinct species earlier referred to genus *Homo*. The segments of the homonoid group for which the names *Homo heidelbergensis* and *H. neanderthalensis* were formerly used are now demoted to the rank of extinct varieties within the existing species: *Homo sapiens* var. *heidelbergensis* and *H. sapiens* var. *neanderthalensis*. Modern man becomes *Homo sapiens* var. *sapiens,* compounding our self-aggrandizement by doubling the number of times we label ourselves as wise.

Such changes in classification and terminology emphasize the continuity of the stream of life and of the evolutionary processes constantly changing its components. They also drive home the fact that a species is in reality a segment of that continuous stream, artificially and somewhat arbitrarily selected for the convenience of men in communicating with each other. In the attempt to think of a species as a natural rather than a man-defined group of individuals, a species is sometimes defined as including all creatures that freely interbreed and excluding all others. This helps the zoologist and botanist on many occasions, but there is a catch in the adverb "freely." Do rare instances of hybrid paternity qualify in spite of the general rule? And what do you do with constitutionally sterile individuals? Obviously, however, this test can hardly be applied to creatures who died many thousand or million years ago and are known only by fossil bones or teeth, or by the weapons, tools, pottery, and other artifacts left at their camp sites or scattered across their hunting grounds. The best we can do is to recognize the taxonomic scheme as reflecting the consensus of men whom we believe to be qualified to have relevant opinions and to consider it as indicative of the significant relationships between individuals separated from each other in space or time.

Chronologic overlap of australopithecines and members of genus *Homo* was definitely established by Leakey[7] in the early 1960s when he found in Kenya the fossils of a hominid to which he gave the name *Homo habilis* (the "handy man") because of the association with crudely fashioned artifacts. Depending on one of the radioactive timekeepers, he dated those fossils at 1.75 million years old, but that date is not unanimously accepted. Even so, *H. habilis* is probably the earliest known representative of the genus.

When asked how old is man, I cannot answer until I know what the questioner means by *man:* the variety of *Homo sapiens* to which we belong; the species as a whole, including its extinct species; or the hominid family, including its extinct genera? My own predilection is to use *man* for all varieties of *Homo sapiens, mankind* for all members of genus *Homo,* and *subhuman hominids* for earlier and now extinct genera and species of the hominid family (although in other contexts I may refer to all existing human beings as *mankind*).

Using that terminology, the record of the emergence of man may be briefly summarized. Subhuman hominids lived in Africa for several million years, prior to about one million years ago. There they were in direct and critical competition with the ancestors of anthropoid apes, old-world monkeys, and predatory felines. It was a down-to-earth struggle for existence. Those that survived were doubtless the best killers; they were also the ones best able to engage in collective and coordinated activities as they perfected their social organization in packs and clans. Appearing about two million years ago, or a little less than that, in eastern and northern Africa, mankind spread to South Africa and to the Eurasian continent. There the earliest records are found in Java; they are correlated with the first interglacial stage of the Pleistocene Epoch and date back about a million years ago. (The Great Ice Age comprises four glacial and three interglacial stages). The mankind fossils found near Peking, China, are correlated with the second interglacial stage and are therefore about a half-million years old. By that time Heidelberg man was living in what is now Germany. It was, however, the Neanderthalers who left the most extensive record of early man. They lived throughout the third interglacial stage and into the fourth glacial stage, an interval of at least 150,000 years, during which time they spread along the shores of the Mediterranean Sea and across the Eurasian continent from the Atlantic to the Pacific Oceans. Their cultural development is shown in the progressive improvement of their stone tools and weapons, the artifacts they fashioned from the bones of slain animals, and the fact that some of them buried their dead. Modern man (*Homo sapiens* var. *sapiens*) first appears in the record shortly before the close of the third interglacial stage, fifty or sixty thousand years ago. The best-known type, Cro-Magnon man, entered western Europe between 42,000 B.C. and 28,000 B.C., displacing or absorbing its earlier hominid populations.

There is today more than a suspicion that the Cro-Magnon Man, unquestionably a member of the species *Homo sapiens* var. *sapiens*, did not actually exterminate Neanderthal Man when he invaded the Dordogne Valley and other parts of western Europe—a region that had been inhabited for the preceding one hundred and fifty thousand years by the Neanderthalers. Instead, as usual during any massive invasion in war or peace, the newcomers may well have interbred with the remnants of the local inhabitants. If that was the case, the relatively new nomenclature is fully justified. More importantly, it would mean that some of the genes from the neanderthal gene pool are still potentially active in the chromosomes of living men. To call a member of the opposing political party or an opponent in a discussion or debate a neanderthal man may possibly have a scientific basis as well as an oratorical impact!

All these various types of mankind and of man continued the competition with other animals that has been noted for their australopithecine ancestors. They were hunters and gatherers of food. Not until ten or twelve thousand years ago, notably in Asia Minor, did any of the food-gatherers become food-producers to any significant extent. There must also have been considerable competition between various bands and clans as each staked out its own territory to be defended against any and all intruders. Continuing improvement in the fashioning of tools and weapons had its obvious survival value. The acquisition of techniques for using fire was undoubtedly of paramount value in the continuing struggle for existence. So also, and perhaps even more importantly, was improvement in the fine art of cooperation. Coordinated activities of individuals within any species of animal, regardless of the class to which the species belongs, depend upon some effective means of communication between those individuals. Among the mammals, communication is largely vocal. So it must have been among our pithecine and hominoid ancestors. One of the distinctive human characteristics is the broad, flat, flexible tongue which permits a much greater variety of sounds to issue from the mouth than can be emitted by any other creature. To provide room from such a tongue, the lower jaw must have a considerable space below the tooth plane, and the two sides of the jaw must meet in front in a broad curve rather than an acute angle. Thus developed the distinctively angular human chin in contrast to the receding chin of subhuman hominoids and other primates. The inference is obvious that instead of indicating strength, determination, and independence of character, as sometimes suggested, a powerful jutting chin really indicates a potentially more vociferous individual.

But improvement of cooperative behavior within a clan or tribe required greater intellectual ability as well as more effective communication between individuals. As the brain, and especially its frontal lobes, became larger, the forehead expanded upward. Together, the development of jaw and forehead eventually produced the characteristic human face in contrast to the simian countenance. As the hominid brain became larger and more capable, its functions expanded. To the ancient office of remembering experiences and observations, with the ability to retrieve needed items from its storehouse, were added the functions of thinking rationally and eventually abstractly, of designing patterns for things and for societal structures, and of becoming vividly aware not only of the physical and biological factors in the environment but of nonmaterial realities as well. The drawings and paintings in the caves of Lascaux in

France and Altamira in Spain, dating from twenty to thirty thousand years ago, as well as the carved figurines found in anthropologists' digs, are especially significant parts of the record of evolution of primitive man.

Thus far we have been concerned almost exclusively with the progressive achievement of the human type of body; that is to say, with man's biological evolution. Even more important is the development of human societies, the organization of individuals in coordinated groups, the flowering of the arts and sciences, the achievement of visible interpersonal relations among individuals, the historic progress toward effective religious ideas and ideals; these are the components of man's cultural evolution.

Somewhere, far back in our ancestral lineage, must have lived the individuals who first flaked a flint to a cutting edge. Probably there were many of these innovators living more or less contemporaneously, each in the isolated territory of his own family or clan. Be that as it may, this initial application of the creative ability of the human artisan demonstrated the fact that *Homo sapiens*, man the thinker, is also *Homo faber*, man the maker. Indeed, the human ability to manufacture implements, tools, weapons, mechanical devices, artifacts in infinite variety, is the most diagnostic characteristic for distinguishing man from all other creatures. Some of us think that Linnaeus, who gave the scientific designation *Homo sapiens* to man in his universal classification of animals and plants in 1735, would have been more sapient had he named man *Homo faber* instead.

Handmade weapons and other implements went far toward compensating the deficiencies of hominid anatomy in the competition that must have been waged with contemporary carnivorous mammals in the early dawn of human history. Only by their invention and use could primitive man have survived among creatures more agile than he, possessing sharp claws and lethal fangs. But these alone could not possibly account for the success of primitive man in subduing the beasts of the field and forest. Organization of groups in which the activities of each individual were coordinated with those of others was an inescapable requisite for continuing existence. Mankind's gradual rise to dominance among terrestrial creatures can be explained only in terms of increasing skill in providing mutual aid and organizing group activities.

Ability to organize for mutual assistance is just as much a part of the human heritage from prehuman ancestors as the form and structure of the body. The important role played by mutual aid in organic evolution has long been recognized by students of geologic life development.[8] Its foregleams can be glimpsed among the records of both invertebrate and vertebrate animals as far back as the Paleozoic Era. Survival of the relatively weak and puny mammals of the Mesozoic Era must be attributed in part to their affection for their offspring and their tender care of their young. Especially in the strain that led to mankind during the Cenozoic Era there must have been an increasing development of habits of cooperation and organized group activities.

Much is being learned today concerning the internal organization of congregations of gregarious wild animals such as herds of antelope, prides of lions, packs of wolves, companies of baboons. Animals usually behave quite differently in the wild than in the zoo or when domesticated. Information about the publicized pecking order among barnyard fowl is not nearly so relevant as knowledge concerning parental care of young and the functions of leadership and territorial rights among chimpanzees in the forest.

Our pithecine ancestors lived in jungles and on grassy steppes. They obeyed the law of the jungle. Many of their mental habits and emotional traits were inherited by the

men of the Old Stone Age. They, too, were killers; most
of the tools recovered by cultural anthropologists from
early archeological sites are weapons. And men who display Old Stone Age patterns of behavior are still with us
today. Fortunately, we know a great deal more about the
law of the jungle today than did Kipling when he equated
it with the primacy of "fang and claw." Robert Ardrey, for
example, has described it in his vivid prose[9] as a combi-
nation of "enmity-amity"—enmity toward those outside
of one's own congregation, amity toward those within it.

The history of mankind during the last half-million years
has been marked by increasingly efficient organization of
individuals in social groups on an amicable basis and by
progressive expansion of the territory within which amity
is sovereign. Familes banded together into clans, clans
joined to form tribes, tribes united to create nations, and
today the dream of a "federation of the world, a parliament
of man" has a much greater possibility of achievement than
ever before. Our heritage from jungle-dwelling ancestors
of the goodly ingredient called amity is a potential vital
force that only needs to be brought to full fruition, liqui-
dating the heritage of enmity as it rises to sovereignty, in
order to realize the prophets' dreams of a terrestrial "City
of God." And in that process it should always be remembered that there must be effective amity within the boundary circle of enmity-amity before that circle can be enlarged
to embrace additional territory.

Even the most primitive organizations among the
emerging hominids provided fertile ground for the cultivation of the seeds of moral principles. Certain kinds of
behavior that might seem desirable to an individual or that
might contribute to his personal welfare were discovered
to be inimical to the welfare of the group as a whole.
Judging from what we now know about group behavior
among chimpanzees and other anthropoid apes, some of

the ancestral subhuman hominids must have been at least
dimly aware of the idea that the individual's responsibility
for the continuing prosperity of the group to which he
belongs transcended even his desire to keep himself alive.
As that idea came to be expressed in signs and words,
codes of approved and disapproved conduct began to be
formulated, tested, revised, and eventually perpetuated in
folk-lore and on tablets. We call them taboos when we
encounter them among primitive tribes, ethical principles
when we discuss them among ourselves.

Somewhere, far back in our ancestral lineage, there must
have been those who first said it would be wrong to do
this, right to do that. In that figurative language of the
Garden of Eden myth in Genesis 2 and 3, such persons
had "partaken of the fruit of the tree of knowledge of good
and evil." And it is more than likely that in at least a few
localities it was a female of the species who first recognized
the necessity for a moral code and then shared her new
insight with her companions. More importantly, knowledge of good and evil is spiritual knowledge, even as the
ability to flake a flint into a tool or weapon requires at least
a modicum of scientific knowledge.

Matters of interpersonal and intergroup relations, of
morals and ethics, are among the cultural aspects of life,
as are also the deeper inner thoughts one has about his
personal relations to the universe and its administration.
The changes these have undergone throughout the history
of mankind constitute a large part of man's cultural evolution as distinguished from his biological evolution.

Man's body, probably even including his brain, is today
more or less stabilized. Certainly we are doing all we can
to standardize it, what with modern systems of education,
progress of public health, beauty contests, athletic com-
petitions, etc., on an international as well as on a local
basis. But the spirit of man is still struggling in travail.
Man's cultural evolution has just begun; its possibilities

for future progress are literally beyond measure. Surely, in this critical moment in cosmic history, when man has just about fulfilled the ancient directive to "be fruitful and multiply and replenish the earth and subdue it," emphasis must be placed upon his further cultural evolution.

Fortunately the tempo of cultural evolution, whether progressive or retrogressive, is much more rapid than that of biological evolution. Changes in organs and structures in the body can be transmitted to offspring only if they are the result of prior changes in the genes, the carriers of inheritable characteristics. Consequently, biological evolution is slow, even in the perspective of geologic time. In contrast, knowledge and emotions, ideas and ideals acquired by one generation may be transmitted immediately and directly to the next generation. It is the function of science not only to extend scientific knowledge but also to screen out from the transmission line any erroneous ideas. It is the function of religion not only to extend spiritual knowledge but also to screen out unworthy ideals and promote the transmission of emotions that bind men together.

For a hundred thousand years or more, the cultural evolution of mankind has been on a provincial basis. Even as late as the beginning of the Christian Era, groups of human beings were more or less isolated from each other by the barriers of seas and oceans, lofty mountain ranges and vast deserts, and the difficulties encountered in travelling great distances. Only within the last two hundred years has the world become one, thanks to modern means of transportation and communication made possible by recent advances in scientific knowledge. The cultural evolution of mankind is now inescapably cosmopolitan in nature. The problems of the contemporary human predicament will never be solved unless we find the way to adjust more competently our on-going cultural evolution to this change from provincialism to cosmopolitanism.

NOTES

1. Elwyn L. Simons, "Early Ape," *Geotimes* 13 (1968): 24.

2. L. S. B. Leakey, "East African fossil Hominoidea and the classification within this superfamily," *Classification and Human Evolution,* Sherwood L. Washburn, ed. (New York: Viking Fund Publications in Anthropology, No. 37, 1963); L. S. B. Leakey, *Olduvai Gorge, 1951–1961* (England: Cambridge University Press, Vol. 1, 1965; Vol. 2, 1967).

3. John T. Robinson, *Early Hominid Posture and Locomotion* (Chicago: University of Chicago Press, 1973). See also, H. M. McHenry's comment on "Australopithecine Anatomy," *Science* 181 (1973: 738–39.

4. Wilfred Le Gros Clark, *Man-apes or Ape-men?* (New York: Holt, Rinehart, and Winston, 1967).

5. Anon., "M C Z Finds, Identifies Hominid Bone," *Harvard University Gazette* 66 (1971): 1, 2.

6. Maitland A. Edey, *The Missing Link* (New York: Time-Life Books, 1972). This is an excellent, popular account of what was known prior to 1972 about the ancestry of mankind. Its publishers distributed to its purchasers in July 1975 a small pamphlet containing a brief description of Lucy, also by Mr. Edey.

7. L. S. B. Leakey, *Illustrated London News* 235 (1959): 219; *Nature* 202 (1964): 7–9.

8. P. Kropotkin, *Mutual Aid, A Factor In Evolution* (New York: Knopf, 1914); W. C. Allee, *Cooperation Among Animals* (New York: Schuman, 1959): Caryl P. Haskins, *Of Societies and Men* (New York: Norton, 1951).

9. Robert Ardrey, *African Genesis* (New York: Atheneum, 1961) 170–74. Ardrey indicates that he derived "enmity-amity" or "amity-enmity" from Herbert Spencer's *The Principles of Ethics,* published in 1892.

6

The Administration
of the Universe*

SCIENCE IS VARIOUSLY designated as a quest for knowledge or as a servant of mankind. The distinction between these two aspects of the scientific enterprise is implicit in every attempt to contrast pure science with applied science. Although valid in theory and sometimes applicable in practice, that distinction is frequently and appropriately blurred in both academic and industrial laboratories. In almost every endeavor to increase human efficiency or comfort, by applying the scientific knowledge already available, it soon becomes necessary to acquire new knowledge. Research and development characteristically go forward hand in hand.

At the deeper level of man's unquenchable curiosity concerning the world in which he lives, the nature and meaning of his own life, and the characteristics of the forces producing the constant changes he observes, the quest for knowledge is itself the servant of mankind. The fundamental concepts of every scientific discipline have something to say about the nature of the universe and the characteristics of its administration. Although it is true that science deals primarily with the immediate or proximate concerns of man rather than with his ultimate concerns,

97

it is nevertheless the duty of science to push our minds as far as they go toward comprehending the nature of the administration of the universe and man's relation to it.

To the best of my recollection, I first heard the rubric, "administration of the universe," from the lips of T. C. Chamberlin during the academic year 1909–1910. He was then the senior professor of geology in the University of Chicago and I was a first-year graduate student in his department. Although in his sixty-seventh year, he was still eight years away from retirement as a member of the faculty. The one formal course of instruction he offered was entitled Principles of Geology. In it he discussed whatever geological problem was currently engaging his research-oriented mind, ranging from the origin of the earth through mountain-making forces to the causes of glacial climates. It was, in fact, a course of study of T. C. Chamberlin, an opportunity to gain insight concerning the workings of his mind, to observe how he communicated his ideas to others, and most importantly to acquire intimate and personal knowledge of a great "scholar, teacher, and gentleman," as his students labelled him. I audited his course that first year in the Graduate School, took it for credit the next year, and audited it again in 1914–1915 when I returned to Chicago to complete the requirements for the Ph.D. degree after three years of teaching at the University of Arkansas. Thus I heard these words drop casually from his lips at least a dozen times and came at last to some comprehension of their meaning in his vocabulary.

For Chamberlin then, and now for me, "the administration of the universe" is a perfectly valid scientific phrase. Like many another, it was coined to reveal some significant knowledge and conceal a considerable amount of ignorance. It simply affirms that the universe is under some kind of administrative regulation, whatever the administrative power may be. It implies only one thing about the nature of the administration: that it is unitary; *administration* not *administrations*. Significantly, administration is not spelled with a capital A in ordinary usage; nor is there any suggestion that *administrator* is an appropriate synonym.

I do not know whether T. C. Chamberlin was the first to express this concept in precisely these words, but the general idea thus conveyed in specific terms is of course an ancient one. It is glimpsed in the thinking of the patriarchal sages of the sixth and fifth centuries B.C. who affirmed, as in the first chapter of the Book of Genesis, that the same power that orders the stars in their courses is also responsible for the presence of man on the earth. It resounds in some of the most majestic verses in the Book of Psalms and is essentially the basis for the philosophical perplexities of Job. It is implicit in all of the far-ranging discussions of natural law, the laws of nature, and the order of nature that have enlivened intelligent discourse throughout many centuries. A universe in which all processes of change operate in accordance with discernible (or potentially discernible) regulations or directives must be subject to some kind of administration.

The evidence that we live in a world of law and order is much more abundantly available to modern man and more authoritatively convincing than ever before. For geologists, the verdict was rendered early in the nineteenth century. In 1795, James Hutton had asserted in his "Theory of the Earth"[1] that in any attempt to explain geological phenomena "chaos and confusion are not to be introduced into the order of nature." Rather, the processes of change now modifying the earth have been operating uniformly throughout its entire history. Even though changes during the lifetime of any one observer may seem slight, when they continue for long periods of time they are sufficient to explain all that we see. Some of Hutton's contemporaries refused to accept this idea, which seems so reasonable to most of us today, and for a few decades serious

controversies persisted between uniformitarians and catastrophists. The truth is, of course, that some of the uniformly operating geologic agents, like the waves and
currents of the sea or the rivers and glaciers of the lands,
can be observed at work at any time and in many places,
whereas others, like the movements in the earth's crust
that cause earthquakes or the eruptions of molten rock
that produce volcanoes, are spasmodic in time and con-
fined to relatively few localities. Long intervals of quies-
cence or impotence extend between shorter episodes of
sometimes catastrophic violence. And what shall we say
about such rare events as the impact of the extraterrestrial
body that produced Meteor Crater in Arizona a couple
thousand years or so ago? Although the catastrophists might
seem to have a point there, we uniformitarians insist that
even impact craters come within the order of nature.

It was Charles Darwin in 1859 who made it possible for
biologists similarly to become uniformitarians. Since his
day, the discovery of the principles of heredity, the genetic
code, the laws of metabolism, the regulatory functions of
endocrine glands, etc., have made it evident that the biological world is just as truly a world of law and order as
is the physical world. The evolution of plants and animals,
including for many of them cultural as well as anatomical
development, is a creative process operating in accordance
with administrative directives amenable to rational analysis.

Research in the broad areas encompassed by the physical and chemical sciences has contributed immeasurably
to human efficiency and comfort in recent years. That research is predicated on the proposition that the transformations of matter and energy with which physicists and
chemists are concerned take place in accordance with rules
and regulations which can be expounded as laws of nature.
One of the most impressive demonstrations that this prop-

osition is valid is found in the periodic table of the elements depicted on the charts now on display in almost every secondary school as well as in all colleges and universities the world around. It is a convincing assembly of credible evidence that the physical universe is organized in conformity with administrative regulations.

The order of the elements in the periodic table is not to be confused with the orderly sequence of letters in the alphabet, although the student may be asked to memorize each at different stages in his academic career. The alphabet was arranged for the convenience of men and women and is a result of enduring custom and continuing agreement. It is manmade, whereas the periodic table was discovered by man. Hydrogen is element number one because the nucleus of a hydrogen atom contains one proton and there is one electron in the space surrounding the nucleus within that atom. Helium is element number two because the helium atom has two protons and two electrons; lithium is number three because its atoms contain three protons and three electrons. And so on through the entire sequence, past uranium with its ninety-two protons and ninety-two electrons to the man produced elements numbered ninety-three to one hundred and two. The numerical arrangement is inherent in the nature of matter made known by scientific research.

The mathematical elegance of the periodic table becomes even more impressive when the several hundred isotopes of the elements, now known, are inserted in it. There is evidently some kind of quantitative relation between the number of protons and neutrons in the atomic nucleus and the number of isotopes of the particular element under investigation. The regulations are such as to set definite limits to the variations in mass number permissible in nature. Add to all this the exquisite precision with which the atoms are now known to change by radioactive decay or in response to high-energy bombardment, and the evidence is conclusive: the material universe is regulated by administrative directives.

There is, to be sure, a large factor of random activity or apparent lawlessness in the individual behavior of such subatomic entities as electrons, mesons, and positrons. The Heisenberg principle of uncertainty comes immediately to mind. It is not yet known, and may never be known, whether this is due to an actual absence of any regulation that might determine such behavior or to the ability of the human mind to discover the pertinent regulation, concealed as it might always be by the obscurity inherent in the limitations of sense perception and epistemology. In either event, it would suggest a characteristic of permissiveness pertaining to the congeries of natural laws that generally seem profoundly obdurate. Be that as it may, whenever and wherever the subatomic entities are organized to form atoms, they display meticulous obedience to regulations that are being spelled out with ever increasing exactness by the physicists and chemists of today.

It is also true that many, perhaps most, of the laws of nature are statistical laws governing the behavior of aggregates of many individuals, each of which seem to act at random. The regulations pertaining to gases and those set forth in explaining the principles of genetics are good examples. But statistical laws are no less binding; they, like other manifestations of the manifold transformations of matter and energy, tell us much about the nature of the administration of the universe.

Basic in the world-view of modern science and the operational research of many scientists today are the fields of force. Best known of these are the gravitational, the geomagnetic, and the electromagnetic fields. The gravitational field is at least a partial answer to the question that Newton left unanswered. The various members of the solar system, for example, appear to be free in empty space; there are no mechanical or material connections between them. How then can each exert a force on the others? Such action at a distance is an essential property of this field, revealing itself as it does in the regulations pertaining to inertia, momentum, and acceleration.

The presence and directives of the geomagnetic field explain the otherwise incomprehensible behavior of the compass needle, one end of which seeks the north magnetic pole unless deterred by local magnetic impulses. It too is characterized by action at a distance; an air-borne magnetometer operates as uniformly as an earth-bound one. Or to go even farther afield, the presence and action of the geomagnetic field account for the trapping of ions in the Van Allen Belts around the earth.

Similarly, the electromagnetic field explains other kinds of action at a distance, such as that involved in electrostatic charges, electromotive force, and the organization of electrons in the various shells within atoms. Some of the distances are minuscule, others are seemingly infinite. The contemporary concept of this field, for example, goes far toward resolving the ancient paradox concerning the nature of light, whether it is a stream of discrete particles or a sequence of wave motions.

A fourth field of force is even now beginning to be recognized. There seems to be a nuclear force field, sovereign within atomic nuclei and responsible for the organization of nucleons to produce ultrafine-scale systems having various degrees of stability, related in part to the complexity of their construction. Doubtless this field will soon be described with greater precision and clarity, now that the concept of force fields is proving so fruitful in contemporary research.

Field is a word with a variety of meanings in common usage, ranging from a potato field or an athletic field to a field of study or a particular specialty within a scientific discipline. Its meaning is usually made clear by the context in which it is used. It is not easy, however, to convey adequately and accurately its meaning within the vocabulary of a modern scientist. Thus d'Abro described a field a score of years ago as "the continuous distribution of some 'condition' prevailing throughout a continuum."[2] Probably a more recent definition by Victor Guillemin is more enlightening: "A region of space in which objects have forces acting on them such as the space around the earth where the gravitational field produces forces on objects proportional to their mass [and to the square of their distance from the earth]. Similarly, electrical fields exert forces on objects proportional to their electric charge. A field may be visualized in terms of imaginary lines of force which point everywhere in the direction of the force and whose density (closeness) is a measure of the strength of the field."[3] No matter which definition is preferred, it would be correct to say that a field, as the word is used in up-to-date scientific context, is an abstract concept rather than a physical reality, although its effects may be and often are physically real. It is neither matter nor energy, yet it may exert command power over both. It cannot be directly perceived by human senses no matter how they are reinforced or extended by the apparatus of scientific laboratories or observatories. Its presence, however, is made known by an abundance of perceptible and measurable events. Should the earth's gravitational field, for example, cease to be, all objects on the earth's surface would instantly fly off into space and the earth itself would promptly become an expanding cloud of molecular particles exploding violently outward at tremendous speeds.

Each of the four fields mentioned in foregoing paragraphs has the capacity to organize and order (or structure) the particular force which characteristics it. These forces may be measured by observing the effects or events they produce. The best known forces, long recognized by physicists, are the gravitational and electromagnetic. "(Magnetic forces arise between electrical charges in motion and are therefore linked with electrical forces in a single term.) Particle research, a phase of recent nuclear investigations, has revealed two additional kinds of forces called simply the *strong force* and the *weak force,* these noncommittal names being indicative of our ignorance concerning their nature."[4] The strong force binds together the protons and neutrons in the nuclei of atoms and organizes them to produce nuclear structures in accordance with the directives of the nuclear field. Like the strong force, the weak force has a very limited range of action, and it too may be a property of the nuclear field. Or there may be two nuclear fields. In any case, it is the electromagnetic field with its electromagnetic force that rivals the gravitational field in the importance of its administrative role in the world of sense perception. The geomagnetic field, although of great significance in the life of man, is only a local and special manifestation of the presumably universal electromagnetic field.

Surprisingly enough, when measured by the parameters used by nuclear physicists, the gravitational force is far weaker than the so-called weak force. It is "the cooperative action of all the particles in a very large object such as the earth [that] adds up to a strong total effect."[5] On the other hand the strong force is actually a hundred times as strong as the electromagnetic force.

Other fields than these four are almost certain to be postulated in the near future. The recognition of one such appears to be just around the corner. Much significant research is now directed toward the discovery of the processes whereby living cells may have evolved from antecedent inorganic chemical compounds in the sterile environment of the lifeless earth far back in Precambrian time three or more billion years ago. Loose macromolecules, similar in composition, form, and structure to cer-

tain of the macromolecules essential to the life processes
of now-living creatures, have been produced in the laboratory by purely chemical synthesis of inorganic substances catalyzed by electrical discharges. But, as George
Gaylord Simpson points out,[6]

> If evolution is to occur and organisms are to progress
> and diversify, still more is necessary. Living things
> must be capable of acquiring new information, of al-
> teration in their stored information, and of its com-
> bination into new but still integrated genetic systems.
> Indeed it now seems that these processes, summed
> up as mutation, recombination, and selection, must
> already be involved in order to get from the stage of
> loose macromolecules to that of true organisms, or
> cellular systems. There must be some kind of feedback
> and encoding leading to increased and diversified ad-
> aptation of the nascent organisms to the available en-
> vironment. Basically such adaptation is the ability to
> reproduce and maintain or increase populations of
> individuals by acquiring, converting, and organizing
> materials and energy available from existing environ-
> ments. These processes of adaptation in populations
> are decidedly different in degree from any involved
> in the prior inorganic synthesis of macromolecules.
> They also seem to be different in kind, but that is
> partly a matter of definition and is also obscured by
> the fact that they must have arisen gradually on the
> basis of properties already present in the organic pre-
> cursors. In any case, something new has definitely
> been added in these stages of the origin of life.

Dr. Simpson goes on to indicate that he does

> not mean to say that material causality has been left
> behind or that some mysterious vitalistic element has

been breathed into the evolving systems. All must still be proceeding without violation of physical and chemical principles. Those principles must, however, now be acting in different ways because they are now involved in holistic, organic, increasingly complex, multimolecular systems that transcend simple chemical bonding.

It is entirely possible, and it seems to me highly probable, that this "acting in different ways" is the response to the directives inherent in an organic field of force, the sovereignty of which is apparent, *vis-a-vis* that of the electromagnetic field, only when simple chemical bonding has produced a physicochemical system of requisite complexity.

Thoughts somewhat akin to these seem to be in the mind of Theodosius Dobshansky,[7] although he does not structure them in terms of field theory. He believes, as do many other biologists, that phenomena on the inorganic, organic, and human levels are subject to different laws peculiar to those levels. Living organisms were of course initially derived from inorganic precursors and mankind from preceding organisms, all in accordance with the regulations inherent in the all-embracing process of evolution. Thus he uses the term "evolutionary transcendence" to denote the new levels of activity displayed in the history of that process.

An organic force field such as suggested here would explain, or help to explain, the transcendence of the organic level *vis-a-vis* the inorganic. Actually it would be no more mysterious in essence than any of the fields of force the presence of which seems to have been established in the physical sciences. Its presence may only be tentatively postulated at present but, as the concepts of field theory spread from the physical sciences into the life sciences, we may expect a reasonably accurate and precise description of it in the near future.

Such a description will presumably go far toward solving one of the obdurate problems inherent in our present knowledge of the process of organic evolution. Geneticists report authoritatively that the mechanisms of heredity—the mutations and recombinations of genes—operate in a seemingly random manner. They guarantee that offspring will be different in greater or lesser degree from their parents, but the differences may be for better or for worse. Yet the record of geologic life development indicates unmistakably that the process of organic evolution produced progressively more capable creatures as it operated in the sequence of time. By what right or virtue has environmental selection decreed that evolution should be progressive? Presumably the directives implicit in the organic field of force are such as to account for this remarkable achievement.

This characteristic of causality, thus attributed to the postulated organic force field, is wholly in keeping with the characteristics of the better known fields of force recognized in the physical sciences. They unmistakably display causality, an attribute not displayed by inorganic material objects perceived directly by the human senses. Moreover, the gravitational, electromagnetic, and nuclear fields are universal if not infinite. The relatively simple systems of the binary stars are gravitationally controlled in precisely the same manner as the more numerous members of the solar system; the myriad stars in distant galaxies are subject to the same gravitational regulations as those in our Milky Way galaxy. The atoms and molecules identified spectroscopically in the most distant stars are identical with those with which we are more familiar in, on, and near the earth. Those fields, furthermore, are durable if not eternal. No matter what may have been the detailed origin of stars and solar systems, several billion years ago, they must have been operating then as now. If the postulated organic field is demonstrated to be real, it too will presumably display these essential characteristics of the known fields.

All of which brings to mind the cryptic statement of the first century Jewish scholar, recorded as the first verse of the Gospel according to John in the Christian Bible: "In the beginning was the Word and the Word was with God and the Word was God." Many and long have been the discussions among theologians, philosophers, and semanticists concerning the meaning of the Greek *logos*, translated thus as *Word*. Goethe, for example, suggested in "Faust" three other German words as substitutes for *Word*, and they have been translated variously as *Mind, Power, Force, Deed*, and *Act*. It will therefore not add much more to the confusion if I suggest a rendering of the passage that might be better comprehended by a modern scientist: In the beginning were the fields, and the fields pertained to the administration of the universe; indeed, the fields *were* and *are* the administration of the universe.

The concept of an organic force field may remind some philosophical biologists of the older concept of an *elan vital* and the metaphysics of the vitalists, ideas that are not today regarded with ardent approval, to say the least. There are similarities, to be sure, but the differences are far greater. An organic force field as here conceptualized is an integral part of the unitary administrative enterprise, taking its appropriate place in an hierarchy of organizing, governing, and directing fields. It is there all the time, not something that is introduced at some stage in cosmic history. Its manifestation in objects and creatures that are amenable to sense perception by human beings has awaited the production of sufficiently complex and appropriately organized systems pertaining to the electromagnetic and other fields. The broadly inclusive concept of an administration of the universe displaying its many-faceted nature through a variety of interdependent force fields implies the essential unit and unbroken continuity of the total world-process. There are no gaps in the material order through which a separate life principle could be inserted. If comparisons with other philosophical models of the universe are to be made, those with the process philosophy of Alfred North Whitehead[8] and Pierre Teilhard de Chardin[9] will reveal the closer resemblances.

But the fields thus far mentioned by no means cover the full range of administrative enterprise. Whatever the directives in the organic force field may prove to be, they must account for the emergence of human nature from the antecedent animal nature of man's progenitors. It is customary nowadays to draw a distinction between the biologic or organic evolution and the cultural or social evolution of mankind. A part of man's cultural evolution may appropriately be designated as his spiritual evolution. This involves the qualitative aspects of human development rather than the quantitative, the intrinsically nonmeasurable factors in the life of man rather than those that are measurable, *values* rather than *objects*.

Although much of human nature indicates merely a great difference in degree between man and other creatures, certain features of man's behavioral patterns seem to represent truly significant differences in kind. Men and women sometimes engage in abstract thinking, such as the construction of a system of non-Euclidean geometry or the postulation of a force field—a kind of conceptual ratiocination in which there is no suggestion that any other creature has ever engaged. Equally distinctive is the awareness of esthetic values which some human beings display. No other creature ever pauses on the hilltop to admire the view or looks with ardent appreciation at a glorious sunset. The creation of works of art—paintings, sculpture, music, poetry, architecture, and so on—the intrinsic value of which is dependent upon or enhanced by their beauty is a uniquely distinctive occupation of human beings. Mankind, and probably mankind alone, responds to esthetic inspiration.

Something similar may be said about man's awareness of ethical principles. The moral and ethical standards of modern man, with their codification in ecclesiastical and civil laws, have apparently evolved from the tribal taboos of primitive races and these in turn from the customs or instincts of territoriality and pecking order implicit in the law of the jungle. If so, there must be something inherent in the process of natural selection that tends toward this kind of response at various stages of evolutionary development in diverse periods of geologic time and under widely varying environmental conditions. One is reminded of the characteristics of durability, universality, and causality attributed to the well-known force fields.

The awareness of ethical and moral principles, with its concomitant sense of personal responsibility, displayed in the behavior of many human beings certainly differs greatly in degree from that detected in the study of nonhuman social groups. It probably also differs in kind. Willingness to lay down one's life, if need be, for an abstract idea or unselfish ideal is on a different level from that on which a creature risks its life to perpetuate the existence of its offspring or its own companions. Man's spiritual aspirations are a part of the broadly inclusive process of organic evolution; they too must be accounted for. Have they not emerged also under the aegis of natural selection?

To paraphrase Dr. Simpson's sagacious commentary on the creation of the first living cells from inorganic antecedents in Precambrian time, something new has definitely been added in the successive stages of the origin and evolution of mankind during Late Tertiary and Quaternary time. All must be proceeding without violation of physical, chemical, and biological principles. Some of those principles must, however, now be acting in different ways because they are involved in the behavior and aspirations

of creatures whose cerebral equipment and psychic potentialities make possible the consciousness of nonmaterial
factors in their environment. It is impeccably logical, and
it seems to me quite necessary, to infer that this acting in
different ways is due to the directives in a spiritual force
field, analogous to the force fields already identified and,
like them, contributing to the total administrative enterprise.

Like the force fields made known by the physical sciences, this postulated spiritual field would be universal,
virtually infinite, and enduring, practically eternal, and
would display the attribute of causality. It, too, must be
intelligible by reason of its pervasiveness and discoverable
by means of the responses made to its directives. These
responses, however, cannot be measured in terms of space
and time; they can be evaluated only in qualitative rather
than quantitative terms. Moreover, this field, like the better known ones, would be bipolar. Analogous to the up
or down and out or in of the gravitational field and to the
positive or negative of the electromagnetic field is the
beautiful or ugly, the ennobling or debasing, the lovely or
hateful, the right or wrong, the good or evil, the amity or
enmity of the spiritual field.

All the fields, generally accepted or tentatively postulated, seem to be operating harmoniously within the
framework of space and time. Each is sovereign in its own
domain and over its appropriate subjects. Their completely
integrated directives are the various aspects of the administration of the law-abiding universe. They constitute the
real environment within which mankind must live.[10]

If this analysis is valid, it follows that the directives of
the various fields, made known by observable responses
to them, provide trustworthy information concerning the
nature of the administration of the universe. Some of the
information thus gained pertains to the administrative regulations for the measurable transformations and transac-

tions of matter and energy. This is scientific knowledge; it is in the public domain; it may be considered as knowledge *about* the administration of the universe. In contrast, some of the information thus gained pertains to the administrative provisions responsible for the esthetic and ethical elements in the universe. This is spiritual knowledge; it is in the private domain; it may be considered as knowledge *of* the administration of the universe. As described in chapter 3, the distinction is analogous to that found in a connoisseur's report concerning a painting, with its description of dimensions, genre, media, and other factual data and the evaluation based fundamentally on his personal response to what he sees.

On both counts, it is clearly evident that the administrative enterprise has been oriented overwhelmingly toward the orderly organization of systems and the integration of systems in supersystems. Electrons, protons, and neutrons are organized to make atoms, many atoms are organized in molecules, some of the molecules are organized as crystals in the rocks of the earth's crust; other molecules are organized to form living cells, some of the cells are organized to produce the multicellular plants and animals, some of the more complexly structured animals are organized in societies running the gamut from hives of bees and hills of ants to schools of fish, herds of elephants, packs of wolves, prides of lions, troops of chimpanzees, and communities of men. The trend toward orderly organization is universal; it characterizes the directives in each of the force fields; it reveals an essential attribute of the administration of the universe.

The record of geologic life development can at best be only fragmentary and many pieces of the jigsaw puzzle still await discovery, but enough is now known about it to permit some rational generalizations concerning the way the processes of natural selection have operated under the aegis of the organic force field. As noted in chapter 4, improvement in organic structures and behavior patterns have come as a result of experimentation, with its concomitant of trial and error. Available raw materials, no matter how inadequte they might seem in retrospect, are used with exquisite ingenuity to produce remarkable results. But the results have not always been of the onward and upward variety. One-celled protozoans have continued to exist successfully from Precambrian times to the present day; in certain lines of descent—among the arthropods, for example,—there has been definite retrogression rather than progress. To put it in biblical terminology, "many are called but few are chosen."

The suggestion of permissiveness in the operations of the organic force field, implicit in my use of directives rather than regulations in referring to its administrative characteristics, is of special significance in any inquiry concerning a possible purpose of life. Obviously it has been the purpose of every species of every plant or animal at every place or time to maintain the existence of its kind of life as long as possible. The achievement of that immediate (local and temporary) goal by the more complexly organized animals involves a learning process or something closely akin to it. The distinctions drawn between instinctive behavior and learned behavior blur toward disappearance when dealing with the responses of intelligent creatures to the directives of the organic and spiritual fields. Learning by experience, with its frequent failures and occasional successes, seems to have played a prominent role in determining the survival of many kinds of life.

But the administrative directive toward orderly organization of increasingly complex systems transcends the urge for survival. The trend of geologic life development has been not only toward creatures with more complicated anatomy and greater awareness of the various constituents of their environment but also toward firmer organization of individuals into groups composed of many members of the same species. A colony of corals is not a social organization, although coral colonies have existed for at least 450 million years, from early Paleozoic time to the present. Among them there is no mutual aid in time of adversity or coordinated activity in quest of food, no assignment of individuals for specific tasks essential to the welfare of the entire group. In contrast, the organization of societies of individuals, characterized by those behavior patterns, increased in firmness among insects and mammals during the Cenozoic Era which began only about 70 million years ago.

Two diverging trends appear in the historic record of social evolution. One is toward a closed society. It culminated in Late Miocene or Early Pliocene time, 10 or 15 million years ago, with certain species of ants, termites, and bees that have persisted practically unchanged to the present day. If mere continuity of existence is the ultimate goal of life, those species have such a start on man that he can never hope to catch up. But I do not think he really wants to. Their perfectly coordinated group activity is a result of regimentation; once a worker bee is born, a worker bee it will remain throughout its entire life; once a warrior ant, always a warrior ant. The other trend in social evolution is toward a free society. Coordination of individual activities is a result of cooperation, not of primordial coercion. Each member of such a society is integrated within its supersystem, not as a cog in a well-meshed, intricate mechanical device, but retaining autonomy, integrity, and freedom to decide whether or not, and how, to participate in contemplated enterprises. This type of orderly social organization is what Abraham Lincoln must have had in mind when he spoke of "government of the people, by the people, and for the people." The historic record of the evolution of *Homo sapiens* and his ancestral lineage over the last several hundred thousand years, including the last few centuries, shows unmistakable progress toward an orderly organization of that kind. This may well prove to be one of the most important elements in the purpose of life, one of the most significant clues to the nature of the administration of the universe.

NOTES

*Revised version of an article first published in *Zygon* 3 (1968) 39–71.

1. James Hutton, "Theory of the Earth," *A Source Book in Geology*, ed. Kirtley F. Mather and Shirley L. Mason (New York: McGraw Hill, 1939) 92–100. Reprinted (New York: Stechert-Hafner, 1964) and (Cambridge University Press, 1970).

2. A. d'Abro, *Rise of the New Physics* (New York: Dover Publications, 1952) 1:71.

3. Victor Guillemin. *The Story of Quantum Mechanics* (New York: Scribner's, 1965), 307.

4. Victor Guillemin, *Quantum Mechanics*, 150.

5. Victor Guillemin, *Quantum Mechanics*, 150.

6. George Gaylord Simpson, The Non-prevalence of Humanoids," *Science* 143 (1964): 769–75; also, chapter 13 in *This View of Life* (New York: Harcourt, Brace and World, 1964).

7. Theodosius Dobzhansky, *The Biology of Ultimate Concern* (New York: New American Library, 1967), chapter 3.

8. Alfred North Whitehead, *Science and the Modern World* (New York: 1925); *Process and Reality* (England: Cambridge University Press, 1929).

9. Pierre Teilhard de Chardin, *The Phenomenon of Man* (New York: Harper 1959); *Man's Place in Nature* (New York: Harper, 1966).

10. cf. F. L. Kunz, "The Reality of the Non-Material," *Main Currents in Modern Thought* (New York: J. Stulman, 1963) 20: 33–42.

7

The Emergence
of Values in
Geologic Life Development*

THE CONCEPT OF SURVIVAL values as an especially
important factor in the process of natural selection was
introduced in chapter 4 and referred to again in chapter
5. In developing the theme indicated by the title of the
present chapter there will be some duplication of the material earlier presented, as I attempt to probe more deeply
into the significance of that concept and thus set the stage
for a consideration of the origins and development of human values. The high ideals and noble aspirations of those
persons whom we regard as humane or truly civilized have
emerged along the path of life through geologic time just
as surely as have the physical characteristics of mankind.
May they not also be a response to directives in what I am
calling the spiritual force field?

Students of organic evolution have long given thought
to survival values in the evolution of all kinds of life, nonhuman as well as human, plants as well as animals. Whenever the word *value* is used, the concept it symbolizes is
oriented toward the future. Survival value is no exception;
indeed the adjective reaffirms the future-orientation of the
noun. Survival values carry a connotation of an objective,
a goal, even a purpose. All living creatures, whether known

only from their fossilized remains or by their presence today, seem to share one common purpose: to maintain as long as possible the continuing existence of their kind of life. This is by no means the equivalent of maintaining the existence of the species to which a creature belongs. When the last of the dinosaurs became extinct, about 70 million years ago, a kind of life that had been maintained for more than 100 million years by countless successive saurian species came to an end. When the three-toed horse became extinct, about 40 million years ago, the kind of life it represented was continued by its lineal descendants through successive equine species to the one-toed horse of Pleistocene and Recent times.

Extinction of a species does not necessarily mean that its survival values were inadequate. A species is only a man-defined segment of what may be a long-continuing sequence of a particular kind of life. Such a sequence of lineally related species and genera may be designated as a *taxon*. In any consideration of survival values, it is the taxon that must be foremost in mind, although the survival value of a species may be temporarily appraised as involving one step on a long road. Such an appraisal is, however, a tricky business. What may seem good for one species may prove fatal for its descendants a few generations or stages later in the taxonomic lineage. Relatively huge bulk may have had great survival value for certain species of saurischian dinosaurs in the Mesozoic Era and for the megatheres among the mammals in early Cenozoic time, but both of *those* taxa were soon, geologically speaking, defunct. The skeletons of many victims of megalomania are strewn in considerable abundance along the path of life.

Survival values have been significantly different for different taxa and for successive species within a taxon at different times. Many of the strains of evolving animals and plants display a cyclical development. The new life

arises in some relatively small geological niche: for example, an embayment of an epicontinental sea at the mar-
gin of a continent for a marine invertebrate fauna, or a
small land area nearly or quite surrounded by epicontinental seas and with its own particular climatic conditions
for a terrestrial vertebrate fauna. In each more or less isolated province the competition for survival leads to the
natural selection of the local champions in terms of their
survival values. Comparative isolation tends toward many
experiments with previously untried organs, structures,
or habits, and favors the development of gene pools that
give viability to the new species and genera. Then comes
one of the far-reaching geographic changes that have occurred so often in earth history. If the marginal embayments are extended to the continental interior by sea
transgression, marine fauna may mingle; if so, the local
champions will be pitted against each other in the continental sweepstakes. Survival values that were adequate
for continuing existence in each of several different provinces are tested under new cosmopolitan conditions. Later
withdrawal of the seas will return the more successful to
marginal provinces, similar to those inhabited by their progenitors in the earlier geologic epoch. This cyclical alternation between provincialism and cosmopolitanism seems
to have played an important role in geologic life development.[1] I will comment later upon its significance with
respect to human values. In a certain sense the cycles are
rhythmic, but they are quite irregular in duration. The
cycles for land animals are out of phase in relation to those
for marine creatures; obviously a time of provincialism for
the inhabitants of marginal epicontinental seas is a time
of cosmopolitanism for the creatures of the land, and vice
versa.

The survival values to which our thoughts have thus far
been directed pertain to organic structures and form and
to the behavior made possible or necessary by those an-

atomical features. They are characteristic of the biology of the species or taxon with which the scientist is concerned. Let me present just one example. The brachiopods are shallow-water, marine, bivalved invertebrates, quite distinct from the clams and oysters which might also be included in that designation. Their fossils are especially abundant in sedimentary rocks throughout the Paleozoic Era, but they constitute only a minor fraction of the marine fauna living today. The great majority of the brachiopod fossils found in strata formed in the first 20 to 30 million years of Early Cambrian time are indicative of inarticulate brachiopods—creatures whose two valves were held together only by the interior muscles, without articulation along a hinge line of projections from one valve into sockets in the other. Even held tightly shut by muscle contraction, the valves could be easily twisted apart by the tentacles of contemporary cephalopods, the most powerful creatures of the Cambrian seas. The cephalopods presumably enjoyed brachiopods on the half shell as an item on their menu even as we prize bluepoints on the half shell today. There was, however, a small minority of articulate brachiopods in some of the Early Cambrian marine embayments. They were beginning to develop interlocking hinges, some of them of considerable length, such that the shell could not be twisted apart without being broken. The survival value of such an apparatus is obvious; by the end of the Cambrian Period (about 100 million years in length) the great majority of brachiopods were articulate. It is but one example of the many episodes in geologic life development during which a small minority possessing superior survival values has become the majority among the creatures of its kind. Any paleontologist can cite scores, if not hundreds, of such events.

The example drawn from brachiopod history pertains to the value of defense mechanisms. There are also many instances of the survival value of organic structures that are useful in aggressive tactics, especially those involving the desire for food. In many of the phylogenetic lineages now known within the more complexly organized branches of the animal kingdom, increased mobility has had obvious survival value, whether the creatures swim freely in water, crawl or creep on the floors of sea or lake, or perambulate on the surface of the land. This involves the ability of the nervous system to coordinate the movements of the various segments of the body of segmented animals and of the paired appendages characteristic of so many orders of animal life.

More significant for our present inquiry is the survival value of an increased awareness of what is going on in their environment that is displayed by successive species in many an evolving taxon. All unicellular protozoa and many of the more lowly multicelled animals are aware only of conditions and things in immediate contact with their cell walls. Cilia and, even better, antennae extend the awareness of creatures possessing such structures to distances of an inch or more from their bodies. Organs of sight, whether merely light-sensitive epidermal cells or single-lensed or compound eyes, had obvious survival value by extending awareness to greater distances. The same is true for organs capable of detecting and identifying sounds or odors. The nature and degree of awareness displayed by any creature that lived in the past or is alive today is probably the best measure of progress as distinguished from mere change. This basis for appraisal is not necessarily equivalent to the measure of an extinct animal's resemblance to man or any other living animal. It simply asks the mathematical question: In how many different ways and to what measurable extent is an animal aware of its surroundings? The answer is found by investigating its anatomy and observing its behavior. This is the least anthropomorphic appraisal of evolutionary achievement we can apply to the various kinds of life we know.

Survival values accruing from anatomical structures may be designated as biological values. In addition there are spiritual values that can be recognized by the natural scientist. One such value is the spirit of adventure that appears to have been influential in the evolution of certain taxa among animals that have achieved a considerable degree of awareness of their surroundings. To accept the spirit of adventure as a reality is to acknowledge that mere continuity of existence is not the only objective of some forms of life. A description of certain events that happened during the Mesozoic Era will clarify and sharpen what I have in mind.

The Mesozoic Era is often called The Age of Reptiles. During almost all of their one hundred and sixty-five million years of existence, dinosaurs were the dominant kinds of terrestrial animals. Some were herbivorous, others carnivorous; they gained long-lasting security for themselves in every habitat afforded by the surface of the land. Even while the saurians were becoming masters of the land, they also deployed into the sea and into the air. In the latter habitat there were two distinct groups of taxa: one, the earlier, included the flying reptiles or pterosaurs that became extinct at the close of the era; the other became the birds that have continued to the present day. We know fairly well how the terrestrial reptiles evolved into flying reptiles and birds, but the question of why this happened can be answered only by speculation. There is no evidence that the land was so overcrowded that some of its inhabitants, ever striving for continuing existence, were forced as a last resort to venture into the radically different and previously untried way of life. Rather, it seems more plausible that there was some kind of urge to launch out into the unknown, to try a new experiment—a spirit of adventure. The venture proved eventually to have survival value: about a hundred million years of continuing existence for pterosaurs and at least fifty million years longer for birds.

Probably the most important of the spiritual values pertains to the spirit of cooperation and mutual aid. This cannot emerge until individuals in a species become organized to form societies. A colony of coral polyps is not a social organization, even though its members live in constant proximity to each other. There is no allocation of specific duties or responsibilities to particular individuals; none spring to the assistance of others whose welfare is endangered or whose lives are threatened; no communication is possible between individuals separated from each other by any appreciable distance. This last-mentioned item means that the minimum requirement for even the most primitive social organism is a nervous system capable of giving its possessor a considerable degree of awareness of its surroundings. It is highly probable that the first animals to attain that capacity were the trilobites. They constitute an extinct class of arthropods, the invertebrate phylum which includes among its many members the modern crustaceans and insects. The trilobites had segmented bodies, paired appendages, compound eyes, antennae, and a well-organized nervous system that must have made them the most intelligent denizens of the Cambrian seas. Their fossils are abundant in the early Paleozoic rocks, reach a climax of diversity near the middle of that era, and disappear completely from the record by its close. Whether or not any of the trilobites developed social organizations will probably never be known. They apparently had the potentialities for doing so and we are free to speculate that the spirit of cooperation may have emerged in them a half-billion years ago.

Be that as it may, we know definitely that this spirit has considerable antiquity—some fifty million years or so. Although certain kinds of insects appear in the record of late

Paleozoic life, the social insects, ants and termites and some bees and wasps, did not arise until much later. Their record begins early in the Tertiary Period and continues on to the present day. It is with the ants that social life has attained its highest expression among insects, and judging by their worldwide distribution the formicine ants display the most efficient social organization for that kind of life. There are fossil ants, preserved in amber and dating back to the Oligocene Epoch, that are scarcely distinguishable from *Formica fusca,* a widely distributed species in Europe and North America today. This is an extraordinary longevity for any complexly structured species. The evidence suggests that the social insects, having climbed to their high state of evolutionary development more than 25 million years ago, have continued to exist on a dead level ever since. Even so, it is mute but conclusive testimony to the survival value of coordinated activity within a societal organization.

Although the ants and termites display the ultimate development of social behavior among insects, the wasps illustrate best the evolution of that way of life.[2] The majority of their species are solitary in habit, others are incipiently social, and still others live in highly organized societies. There can be no doubt that social behavior began among insects with parental care of offspring. Evidently human societies began the same way. As it evolved, individualism was increasingly submerged for the welfare of the collectivized group. A rigid caste system was established, different for termites from that for ants, but equally inflexible for all. Interestingly enough, social habits have arisen among insects no fewer than twenty-four times in as many different groups of solitary insects. Some of these have developed only the rudiments of social behavior, but all tend in the same direction. The caste system involves from three to five castes: the queens whose only function is egglaying; the drones, males performing no function other than reproduction; workers, females whose sex organs are usually undeveloped and who are responsible for the manifold tasks of housekeeping and maintaining the food supply for all in the nest or hive; and warriors, also usually underdeveloped females who defend the nest and occasionally sally forth to enslave workers from other nests or capture larvae to be reared in slavery. In some insect societies there is no separate warrior caste and that function is performed on occasion by the workers. In others there are two castes of workers, each performing special duties. These may run a wide gamut from excavating subterranean rooms and corridors, patrolling the surrounding area to scavenge everything that might serve as food and tending the herds of domesticated aphid "cows," to farming beds of fungus or gathering nectar from flowering plants. This sounds like an exciting variety of occupations that might stimulate resourceful individual behavior. But not so; the inescapable regimentation of this instinctively established way of life has reduced the members of all insect societies to mediocrity of appearance and behavior. Brilliant individualism has vanished. The thrilling slogan of the Three Musketeers has been curtailed to one for all, with no suggestion of all for one, except perchance with reference to the queens.

Societal organization, with its concomitant spirit of cooperation, is commonly displayed also among vertebrate animals other than the more primitive members of this phylum. They all have a spinal cord, and most have some sort of brain and nervous system as well as sense organs of greater or lesser efficiency. This permits a considerable degree of awareness of the various factors in their environments and makes them able to communicate with each other, at least in feeble ways. Among the vertebrate societies, some are loosely organized, others firmly structured. Organization of a group of individuals may be for procuring food, for defense against predators, for attack upon other animals, or for construction of shelters. Coyotes, for example, customarily hunt in packs, and beavers join together to build their dams. In several mammalian taxa it is apparent that survival has depended primarily on the effectiveness of coordinated activities made possible by social organization. Students of evolutionary processes have long recognized the value of the spirit of cooperation and the mutual aid it engenders.

This is especially critical in the phylogenetic lineage that has led to modern man. It was appropriately stressed in chapter 5 where the historical development of that lineage was traced from remote antiquity. The evolution of human anatomy and physiology—man's biological evolution—has evidently been in accordance with the same principles and directives as those to which all other animals are subject. So also has been his development of mutual aid and coordinated activities, a part of man's cultural evolution. In some ways the coordination of activities displayed in the best insect societies might be considered as superior to that observable in the great majority of contemporary human societies. In both taxa the spirit of cooperation has had preeminent survival value.

There is necessarily a close correlation between values and awareness, a term that I prefer to consciousness, although in many contexts the two are synonymous. All human beings are aware of the gravitational field. We share that awareness with many other animals; it is a part of our heritage from ancestral creatures. But our most remote ancestors probably lacked that awareness. They were among the one-celled protozoans of the ancient Precambrian seas. They sought no food, but waited for the waves and currents of the water in which they were immersed to bring them into close contact with particles of food, organic matter synthesized by primitive plants or the protoplasm of other protozoans. It mattered not at all which way was up, which way down, which to the right or to the left.

When, however, those primitive protozoans developed cilia that could propel them, even inefficiently, through the water, and they could go in search of food, those directions established by the gravitational field began to have vital significance. Awareness of that field became more and more apparent, as for example, the Paleozoic fish with their much more efficient organs of propulsion developed the means whereby they had a sense of balance and knew how to propel their bodies in desired directions and keep them in desired attitudes. Even more essential for the survival of their descendants—the terrestrial, aquatic and aerial vertebrates—was their awareness of the gravitational field and the perfecting of the balancing organs and structures in response to it. We human beings customarily respond instinctively to the directives of the gravitational field because of that age-old heritage, long embedded in our genes and chromosomes. On occasion, however, an acrobat trains himself to respond in unusual ways to those directives. And now that astronauts orbit the earth within its gravitational field and are propelled beyond it into that of the moon, we have become far more intellectually aware of this fundamental factor in our environment than any of our ancestors could possibly have been.

In contrast, no creature, living or extinct, is known to have made response to the presence and power of the geomagnetic field until man invented the magnetic compass and the more sophisticated instruments of recent years that depend upon its directives. Some biologists have toyed with the idea that the mysterious navigational ability of such fish as the Alaska salmon returning to ancestral spawning localities and such birds as the homing pigeon and migratory geese involves their possession of a magnetic sense that enables them to distinguish north from south and east from west, in other words, that these particular creatures are aware of the geomagnetic field. But most scientists are justifiably skeptical of this interpretation of these animals' as yet incomprehensible behavior. Certainly there is nothing instinctive about man's use of the compass or other instruments of its kind; this is an art that must be learned by each person who would use them.

On the other hand, awareness of certain properties of the electromagnetic field dates nearly as far back in earth history as does awareness of the earth's gravitational field. Visible light is only a small segment of the widely ranging electromagnetic spectrum, but it has long been of prime importance in the evolution of living creatures. Primordial protozoans were completely unaware of the difference between light and dark although they may possibly have responded to changes in temperature of the water in which they were immersed. Soon, however, some of their multicellular descendants developed light-sensitive cells in the epithelial tissues of their bodies. This gave them an obvious advantage in seeking and remaining in the surface waters of the Precambrian seas where plant food was more abundant than in the dark depths of the ocean. With the advent of much more complexly organized invertebrates and later of vertebrates, efficient eyes became essential to survival. Some are simple, others compound, some give monoscopic vision, others stereoscopic. The eyes of several other kinds of animal are more efficient than the human eye, although our binocular eyes, with their stereoscopic vision, and the nerve system that enables us to interpret what they see, have served well the evolution of the hominoid taxon from the ancestral prehuman species. The range of light visible to man in the electromagnetic spectrum is only from the longer wavelength red light to the shorter wavelength violet light, but there are animals that can see infrared and possibly ultraviolet light to their significant advantage. Only within the last century or two has mankind become aware of the much broader spectrum of the electromagnetic field. Not only have telescopes and microscopes improved manyfold the human ability to see directly, but much more complicated apparatus has been invented to make man aware of the presence in his environment of the invisible portions of that spectrum, ranging upward from infrared light through the wavelengths used by radar, the shortwave and commercial radio bands, to the ultra-long wavelengths of ship-to-ship and ship-to-shore communications, and downward from ultraviolet light through soft and hard X-rays to gamma radiation. Awareness of the electromagnetic field and responses to its forces began in a small way a couple billion years ago and has reached its present, but not necessarily its final, peak of achievement only within the last few years.

Awareness of the directly visible electromagnetic forces and the responses made to them by organisms has had obvious survival value. Can the same be said for the new knowledge about other activities conditioned by the electromagnetic field? Certainly X-ray therapy has saved many lives and radio-communication has made possible the rescue of many shipwrecked mariners. But the question may well be asked as to the survival value of such innovations as the hot line between Washington and the Kremlin. Certainly the men responsible for establishing that means of instant communication had hopes, if not expectations, that it would contribute greatly toward the survival of the U.S. and the U.S.S.R., if not also of mankind in general. Whether or not it will do so, or continue to do so, depends upon the spirit of cooperation and the quality of good will that motivate the men at either end of that line.

Much has been said in chapter 5 and earlier in this chapter about the organization of various animal species in more or less firmly structured societies. The tendency toward coordinated activities of individuals so organized appears in many taxonomic lineages, some of them only distantly related to each other. Societal organization may be observed at many different times in the history of geologic life development, under a great variety of environmental circumstances. It would seem to be the response of many animals to a directive in the postulated organic force field, which like the other fields referred to in chapter 6 is characterized by universality, durability and causality.

Men and women have obviously inherited this tendency toward coordination of individual behavior from their predecessors in man's taxonomic lineage, but they are putting it into practice in ways that transcend anything that has gone before. Voluntary cooperation is apparently a novel way to achieve coordinated activities; goodwill has overtones not heard in mutual aid. The rudiments of the human spirit may have been present in the psychological makeup of some of the prehuman australopithecines, but in a very real sense the spirit of man is his alone. It may well be interpreted as a response to directives in the postulated spiritual force field. How far back in evolutionary history the first response to the directives of that field was made, I do not know. There seems to be no sharp demarkation between such a response and the numerous responses made to the directives in the organic field. The evidence for the continuity of the process of organic evolution is ubiquitous.

The human spirit today is, however, in large measure a product of man's cultural evolution as distinct from his biological evolution. We have doubtless inherited from our forebears certain genes that dictate anatomical patterns potentially favorable for spiritual development. Human values began to emerge along the path of life at the same time that hominid bodies began to display the specific anatomical characteristics of genus *Homo*. It is not by accident that anthropologists designate the collections of artifacts secured from their exploratory sites as cultures and arrange them in sequential stages. Cultural evolution is in fact only the realization of potentials made available by antecedent biological evolution.

For thousands of generations, evolution within the hominoid taxa has been under the influence of provincial conditions, but provincialism has given way to cosmopolitanism during the last few hundred years. This radical modification is the result not of geologic or geographic changes but of human activities. Continuing improvement in means of transportation on sea or land or in the air and in methods of communication has made mankind the most cosmopolitan of all animals. Many of the values and consequent behavior patterns that were adequate for survival under the old provincialism may be quite inadequate under the new cosmopolitanism.

Thanks to science and technology, we live today in a world of potential abundance and inescapable interdependence. For the first time in human history it is now possible to use the rich resources of the earth for the welfare of all mankind. In grasping that opportunity it will be necessary to engage in carefully planned collective action on a scale and in ways that were scarcely imaginable a century ago. Coordination of the activities of the individuals in a society may be accomplished in either of two ways: by coercion, externally applied through political, economic, or social pressure, with brute force and the threat of imprisonment or death, if necessary; or by cooperation, internally stimulated by education and persuasion and freely and intelligently given. If we choose the first-mentioned of these two ways, the future of human cultural evolution will parallel the evolution of the social insects in the past. It is an experiment already tried and found wanting; social insects have existed on a dead level for at least ten million years. If we choose the second-mentioned way, we will be engaged in an experiment that seems never yet to have been tried.

It is however an experiment that has great appeal to many of us in spite of the disappointments and frustrations of these mid-century years. Actually there is much in its favor. As Wheeler[2] pointed out many years ago, insect

societies represent final and relatively stable accomplishments which have developed along purely physiological
and instinctive lines. This instinctive basis, with consequent absence of education and cultural tradition, consti-
tutes a fundamental difference between them and the human
societies. The cultural evolution of modern man reaches
into everything involved in the organization of human
societies and in the endeavor to resolve the paradox of the
individual and his social organization in ways that will
enhance his unique personality.

The most important among the distinctively human values that should be cultivated today among all populations
are those pertaining to the integrity, the freedoms, and
the responsibilities of each member of the rapidly changing
sociopolitical organization of mankind. Such values have
emerged along the path of life only during the last few
thousand years. Awareness of them had to wait for the
evolutionary development of a creature such as *Homo sapiens* with his unique genetic capabilities of mind and body.
Fortunately, competent geneticists affirm that the gene pool
of existing populations is adequate to produce human beings
who are aware of the values in life that are essential to the
attainment of a truly humane civilization.[3]

NOTES

*Based in part on a paper with this title presented at a con-
ference on "Human Values and Natural Science" under the auspices of the State University College in Geneseo, New York, and
first published in *Zygon* 4 (1969): 12–23.

1. Kirtley F. Mather, "Geologic Factors in Organic Evolution,"
Ohio Journal of Science 24 (1924): 117–45.

2. W. M. Wheeler, *The Social Insects, Their Origin and Evolution*
(New York: Harcourt, Brace and Co., 1928).

3. cf. Rene Dubos, *So Human an Animal* (New York: Charles
Scribner's Sons, 1968), paperback edition, 192.

The Future
of Mankind as
Inhabitants of the Earth

ALTHOUGH IT MAY BE possible for mankind to guide and direct its own evolution in accordance with the directives of the permissive universe, many would argue that the opportunity to do so will not be available. Various disaster theories have been postulated in recent years, along with the threat of nuclear disaster, that cast doubt on the survival of human beings. In contrast to these theories, I argue that the future is wide open for mankind. All available evidence combines to lead us to the confident expectation that the earth will continue to be a comfortably habitable abode for creatures like ourselves for many scores, if not for many hundreds, of millions of years to come. In order for this to happen, however, mankind must solve problems related to climate, nuclear weapons, resource management, and population pressures.

Surface temperatures of the earth, the most important item in any consideration of its long-range habitability, are determined by the receipt of solar energy distributed through atmospheric agencies and the forces operating in the geomagnetic field. For any given area of land the contribution of warmth from the earth's hot interior in a year is just about equal to the heat received from the sun in twenty minutes by an equal area in equatorial latitudes under a clear sky at midday. The nineteenth century picture of an earth initially fiery hot but progressively cooling so that yesterday it displayed a glacial climate and tomorrow it will be too frigid to support life anywhere has long since been thrown into the discard. The earth will grow old and die only as a result of failure to receive adequate supplies of radiant energy from the sun.

The future habitability of the earth will therefore be determined by the ability of the sun to continue its discharge of life-supporting energy at approximately the same rate as it is doing today. Nearly a century ago, when Lord Kelvin computed that the sun had only about a hundred million years' supply of that energy, it was generally believed that the sun was only a storehouse of energy. But in the last few decades, the nuclear physicists have enabled the astrophysicists to change that picture completely. The sun is a furnace, producing heat and other forms of energy about as rapidly as they are dissipated in surrounding space. The fuel for that furnace is hydrogen; heat is produced by the fusion of hydrogen atoms to form helium atoms and release energy. It is the same process as that which powers the hydrogen bomb. The question is not how long the sun's store of energy will last, but how long its present supply of fuel will keep the furnace burning. The answer is in the order of magnitude of four billion years!

Nor is there any likelihood that the space relations between earth and sun will change appreciably within scores of millions of years and put the earth either too close to the sun or too distant from it for comfort. Moreover, the lurid pictures of a sudden catastrophic debacle resulting from collision with some other heavenly body—comet, planet, star, or what you will—are products of a vivid imagination wholly without foundation in astronomic fact or theory. The gravitational field is durable if not infinite, and gravitational forces act in a trustworthy, uniform manner as a part of the continuing administrative procedures.

The geologist may, therefore, turn with confidence from the long perspective of the geologic past with its $3^{1}/_{2}$ to 4 billion years of recorded earth history to a similarly long prospect for the future.

It should not be inferred, however, that the earth will continue in the future to display the same environmental conditions as those which we enjoy today. The history of mankind thus far has been enacted against a background that in the full perspective of earth history is truly extraordinary. The geologic period in which we live is a time of unusually rugged and extensive lands, with notably varied climate ranging from the glacial cold of Greenland and Antarctica to the oppressive warmth and humidity of certain equatorial regions. Such conditions have apparently recurred several times at long-spaced intervals since the oldest known rocks were formed, but added together, the time thus represented cannot be as much as a fourth of geologic time. Much more characteristic of earth history as a whole have been the conditions illustrated by those periods when corals thrived in shallow seas occupying the site of Baffin Land and North Greenland, and coal-forming plants flourished on Antarctica. The probability is strong that eventually, say in 5 or 10 million years, the earth will display again the physical conditions of many past geologic periods that were characterized by broad low lands, wide shallow seas, and uniform genial climate.

But most of us have a greater interest in the next few centuries than in the subsequent millions of years. Minor changes in climate will doubtless occur just as they have in the last few thousand years. Unfortunately, or perhaps fortunately, there is no basis for prediction concerning their nature, whether for better or for worse. There is really no good reason for referring to the present as a postglacial epoch; it may prove to be an interglacial epoch. But our ancestors weathered ice ages in the past, and presumably we are better equipped for such contingencies than they were. Should the average annual temperature of the earth as a whole be reduced something like 10°F. and remain at that lower level for a few millenia, it is likely that the greater part of Canada, the northern United States, and the Scandinavian countries would again be buried beneath great ice sheets. But in consequence of the removal of water from the sea as vapor to form the snow to produce the glacial ice, considerable areas now shallowly submerged along the coast lines in middle and equatorial latitudes would emerge as dry land. Indeed, it is likely that the area of land suitable for human abode would be nearly or quite as great at the climax of a glacial period as it is today.

By the same token, the disappearance of existing bodies of glacial ice as a result of rapid amelioration of climate in the not-distant future would, if it occurred, be a decidedly mixed blessing. If the water now imprisoned in the ice on Greenland and Antarctica were returned to the sea without any compensating changes in crustal elevation, sea level would be raised 150 to 200 feet the world around. Considering the number of people who now work or sleep in buildings in metropolitan communities not over 150 feet above sea level, the importance of such a change is readily apparent. It is an interesting pastime to speculate what the citizens of New York City, for example, would do if the level of the North Atlantic Ocean should begin to rise at a rate of one or two inches per year and continue that rise for several hundred years, a prospect quite within rational possibilities. Would they build ever-higher dikes around their densely populated islands and continue to operate the marts of trade and the fun palaces below sea level? Or would they move the city en masse to higher ground in Westchester County or northwest New Jersey?

From the geologist's point of view, such changes in topography, geography, and climate as are likely to take place in the next few thousand years are relatively trivial matters. With due deference to their nature and effects, there is nothing to be expected from such sources that would seriously deter the human species from continuing a reasonably comfortable existence on the surface of the earth for an indefinitely long period of time, a period to be measured in millions rather than mere thousands of years.

Actually man himself is more likely to render the terrestrial environment uninhabitable for his kind of life than are the natural changes that will occur in the next few centuries. One thinks immediately of the possibilities of a nuclear holocaust whereby mankind would go out with a bang. That is undoubtedly a distinct possibility. The overkill capacity of the nuclear explosives now in the stockpiles of the departments of defense of the United States and the Soviet Union provides adequate means whereby mankind the world around could commit collective suicide. But to accomplish that feat would require careful advance planning and the prior organization of men, machines, and materials on an unprecedented scale, planning and organization that, to the best of my knowledge, is not being considered anywhere. It is possible World War III would not do the job, although of course it would be a nuclear war. Most of the nuclear warheads would impact and most of the lethal fall-out from the mushroom clouds would descend within the northern hemisphere. Vast areas of the United States, Canada, Western Europe and the Soviet Union would become a lifeless wasteland. If, by the time of that insane war, China had acquired greater and more effective nuclear capability, the wasteland would stretch across much more of Asia, but even so there would be many extensive areas untouched by warheads, especially in the southern hemisphere. Interchange of air masses between the two hemispheres, at least in the lower part of the atmosphere within twenty-five or thirty miles of the earth's surface, is almost negligible, and by the time the radioactive gases and particles had been carried across equatorial regions in the upper atmosphere, most of their lethal components would have lost their potency. No one knows how many of the earth's inhabitants would be killed or maimed by World War III. There might still be millions of human beings left untouched, and many of them would be highly intelligent, well-educated persons quite capable of carrying on. Winston Churchill may have exaggerated when he was contemplating the consequences of nuclear warfare and said "the Stone Age may return on the gleaming wings of Science." If mankind really wants to commit collective suicide and destroy all civilization, the location of nuclear explosions, the ballastic missile targets and the bomber missions must be specifically planned to form a world-wide network spaced with due deference to innumerable details of movements of airmasses; then arrangements must be made to detonate all the explosives within a period of two or three weeks. That is why I said it is a job that World War III might not do.

This optimistic (?) conclusion, however, leaves me just as unhappy as it does you. I recall a remark attributed to a provincially minded member of the United States Senate who was arguing for the immediate deployment of anti-ballistic-missile missiles, the so-called ABM system, to protect the second-strike capability of the missiles already emplaced at various strategic places in the United States. His argument was to the effect that "if nuclear war reduced the human race to a new Adam and Eve," he wanted them to be Americans. I fear the senator would be disappointed; the builders of a new, postnuclear war world are much more likely to be the citizens of an African or Latin-American nation, Australians, New Zealanders, or possibly Japanese. There may be a few Americans around to serve as consultants, but the principal responsibility for continuing the evolution of mankind despite the temporary disaster would rest on other nationalities.

We may as well accept the fact that there is today no adequate defense against an all-out nuclear attack, nor is such a defense just around the corner in any corridor of any research and development institution. To be quite frank, it is the balance of terror that has maintained such peace among nations as we have had since World War II. In military parlance, the terror is second-strike capability, and it is to maintain the credibility of that deterrent of a first strike that both the United States and the Soviet Union now have their overkill capacity. But the balance of terror is frighteningly precarious; to maintain it inevitably requires a continuation of the arms race, with all its obviously harmful consequences and its impossibility of ever achieving a stable conclusion. I need say no more here about the utter foolishness of such plans for man's future.

Instead, I recommend that we note with humility the fact that the United States is the only nation thus far which has hurled nuclear warheads upon a foe, observe with gratitude the discretion here and abroad that has given us a quarter-century respite from nuclear warfare, rejoice that comprehension of both the good and evil consequences of the use of nuclear reactions is now widely spread in all parts of the world, and proceed to consider the future of man in the expectation that it will not be abruptly terminated or severely disturbed by a nuclear holocaust.

Returning then, to the statement with which this chapter started: the future may be wide open for mankind, but wide is a spatial adjective and it implies boundaries. The permission to continue indefinitely as inhabitants of the earth is bounded by inflexible limitations. Some of these are inherent in the administrative regulations and directives by which the cosmos is kept in order and the processes of organic evolution continue in operation. Others are the result of man's inheritance from his predecessors in the procession of the living. All must be given careful thought if man's future is to be intelligently surveyed and charted.

It is sometimes said that science enables man to conquer nature, but this is at best only a half-truth. The whole truth is found in the statement that science enables men and women to use more effectively, for whatever purposes they choose, the materials and energies provided by nature in their terrestrial environment. The future of mankind is limited by the characteristics, quantities, and availabilities of those materials and energies. The swift speed of man's cultural evolution during the last two centuries, with its establishment of industrialized civilizations in many parts of the world, has involved an unprecedented consumption of those natural resources. During the fifty years between 1920 and 1970, for example, more metallic ores and mineral fuels were mined, processed, and consumed throughout the world than in all the preceding history of mankind. The question is pertinent and pressing: for how many years in the future will the abundant but limited resources of the earth be adequate to sustain man's burgeoning technologic culture?

In any attempt to answer that question, a distinction must be drawn between the two categories into which each specific resource should be assigned: some resources are renewable and may be considered as man's annual income; others are nonrenewable and should be considered as nature's stored capital. Preeminent among the renewable resources is waterpower. Men have harnessed about two million of the six million horsepower that formerly ran wild at Niagara Falls. That energy is used each year in the form of hydroelectricity, with no fear that it will be used up. As long as the earth continues to circle the sun at a respectful distance, water will continue to run downhill and waterpower will be available to produce electrical energy with which to turn the wheels of industry and commerce and to heat and light our homes, factories, and offices.

In contrast, when we use the mineral fuels—coal, petroleum, natural gas, and uranium, or the metallic ores, iron, copper, lead, zinc, aluminum, and so forth—we use them up. They are nonrenewable resources. It is of course true that the same geologic processes responsible for their presence in the earth's crust are continuing to operate today as in the past. But in comparison with the feverish haste of man's insatiable demands, the creative processes in nature's laboratory operate so slowly that, for all practical purposes, our planet must be reckoned as a storehouse of mineral wealth, not as a factory in which that wealth is being generated year by year. It is as though Old Mother Earth has a cupboard richly stocked with a vast amount and a great variety of goods essential for human existence in this age of science and technology. Some of the packages of goods are readily discernible; others are well-concealed on the back shelves. Each year we go to the cupboard and take down a few packages of the goods stored therein; if we keep going long enough, some day somebody will find that the cupboard is bare. Therefore, such resources must be treated as capital assets rather than as expendable income.

Hydroelectricity is by far the most extensively used of all the renewable sources of energy. Even so, it contributes less than a third of the electrical energy now being used throughout the world. Almost all the rest is thermoelectricity produced by the combustion of coal, petroleum, and natural gas or by the heat of nuclear reactions. In the United States, the amount of hydroelectricity made available by harnessing waterfalls and damming rivers has steadily increased during the last forty years, but the percentage of its contribution to the rising total of electrical energy has steadily decreased.

Even so, it is pertinent when looking to the future to raise the question whether it would be possible to meet mankind's needs for electrical energy solely from this renewable resource. If all the potential hydroelectric installations, the world around, could be developed, would the energy thus made available be adequate so that man could live within his annual income? The answer is almost certainly in the negative. No precise quantitative estimates of potential waterpower resources can be made. Economic factors of capital investment, annual upkeep, and value of output are just as important as the physical geography, hydrology, and meteorology of the site under consideration. The latter will be reasonably stable for some time to come, but the former are subject to rapid changes. Dealing then with orders of magnitude rather than with precise figures, it would appear that the total amount of energy derivable from falling water is not much greater than the energy from all sources in actual use in the late 1960s. Only a meager surplus, if any, is available to meet the great increase in demand resulting from imminent industrialization of less developed regions and the rapid increase in population. Incidentally, nearly 50 percent of the potential waterpower resources of the world is in Africa, less than 15 percent in North America. Fortunately, however, there are other renewable energy resources available for man's use when needed. These include the radiant energy of the sun, concentrated by mirrors and lenses, the vast power of the tides along the seashore, the wind, the heat stored or generated within the earth's crust, and the temperature gradients within the water of the oceans. Efficient techniques are now known for the utilization of energy from each of these sources, and significant amounts of power are even now being used at various places from each of them except the last one on the list. Only the adverse economic factors prevent greater use of them at a time when nonrenewable sources of energy, such as coal and petroleum, are still abundantly available.

The energy produced in nuclear reactors must also be considered in this context. A small but significant fraction of the electricity consumed in several countries now comes from this source, the self-sustaining chain reaction of atomic fission under controlled conditions. The fissionable materials now in use are predominantly uranium-235 and plutonium-239. Both are derived from uranium ore minerals and are therefore to be considered, at least theoretically, as nonrenewable resources, like the other mineral fuels. The known world reserves of uranium ore, exploitable by mining methods now in general use, are, however, capable of yielding at least ten and probably twenty times as much energy as would be produced by the combustion of the world's total supply of petroleum, natural gas, and coal. That is so large a figure in relation to the world's annual consumption of energy that one is tempted to think of nuclear energy as a part of man's potential annual income with no fear that it might be used up in the foreseeable future.

There are, however, three obdurate problems that must be solved before electricity is produced in large amounts by means of nuclear reactors. One is a physical problem stemming from the fact that nuclear reactors, operating on a scale adequate to yield significant amounts of energy, inevitably produce amounts of highly radioactive waste materials which could easily pollute the environment in a manner and to an extent that are far beyond any acceptable limits. No one has yet come up with a truly satisfactory procedure for the safe disposal of such waste. The second problem arises from the possibility of human error in the design and operation of nuclear power plants. Nuclear reactors are dangerous devices; a major accident would have catastrophic effects. Much has been done to ensure the safety of their operation and to date there has been no major nuclear mishap involving injuries in any American power plant. But many knowledgeable technicians and members of responsible supervisory boards believe that the problem of reactor safety has been only partly solved. There must be additional research and further experience concerning known weaknesses in quality-control practices, design defects, equipment malfunctions, and operator errors before the risks of operating nuclear reactors are reduced to an acceptable level. The third problem stems from the fact that fissionable material used to produce useful energy may also be used to produce weapons similar to the atomic bombs that devastated Hiroshima and Nagasaki. Are there, or can there ever be, safeguards adequate to prevent the theft or diversion of the plutonium-239 or uranium-235 from nuclear power plants by terrorists, guerrillas, or irresponsible governments in whose hands the weapons thus made available could be used for blackmail or combat? Every increase in the number of nuclear reactors and in their distribution throughout the world intensifies that problem of international security.

Certainly there should be no crash program for the proliferation of nuclear reactors to meet the needs of energy-hungry humanity until all three of those problems have been much more satisfactorily resolved than they are today. Fortunately, there are alternative sources of energy, unquestionably in the category of renewable resources, on which greater emphasis should be placed.

In the meantime, the use of small-scale reactors, designed for the production of radioactive substances to be used in medical diagnosis, treatment, and research, should be encouraged. Such reactors are far less hazardous than those designed for power production and there seems to be no adequate alternative method for obtaining their highly valuable products.

Research should be continued, moreover, concerning

the production of useful energy by means of nuclear fusion. When hydrogen bombs, with their awesome megaton explosive power, were first perfected in the 1950s, hopes were high that this kind of thermonuclear chain reaction could be brought under control to yield energy for peaceful purposes. If that could be done, man's annual income of mechanical and electrical energy would easily meet all requirements for eons to come. Unfortunately, those high hopes have been dashed in subsequent years by the failure of well-planned and adequately financed research to discover ways and means for taming the thermonuclear reaction. That research has revealed much valuable information about plasmas, lasers, and electromagnetic forces, but at this time it would be only wishful thinking to include thermonuclear energy among our assets for the near future.

The final quarter of the twentieth century will presumably be marked by the transition from our previous major reliance for energy on nonrenewable resources (coal, petroleum, natural gas) to our future major reliance on renewable resources (waterpower, geothermal energy, solar radiation, nuclear fission). The order in which those four renewable sources of energy are listed indicates my forecast of the relative contribution of each to mankind's energy budget during the twenty-first century, decreasing from waterpower to nuclear fission. That forecast is of course tinctured, but I hope not tainted, by my lifelong devotion to geology. Be that as it may, any long-range energy policy to which the United States, for example, commits itself should include extensive research and development directed toward prompt expansion of the use of energy derived from geothermal sources and solar radiation.

Summarizing this aspect of the relationship between mankind and the terrestrial environment, it may be stated with confidence that as a result of the technologic application of scientific knowledge it is now entirely possible for human beings to live within their annual income from renewable energy resources. Mankind's future is assured, insofar as requirements for mechanical and electrical energy are concerned.

Is there similar assurance concerning mankind's requirements for food? Human sustenance is derived from the products of farms and grasslands, the forests and tundras, the living organisms in seas, lakes, and rivers. Foods produced by cultivation of the soil—cereals, herbs, fruits and nuts, and the eggs, milk, and meat from domesticated animals—are predominant; the harvest from seas and lakes is next in importance. Both are now, at least potentially, in the category of renewable resources.

Prior to the twentieth century that could not have been said for the products of the soil, but it is true today. Soil fertility depends upon the presence of adequate amounts of water-soluble compounds containing such elements as nitrogen, phosphorus, potassium, and calcium. These are the essential ingredients of all fertilizers, whether of plant or animal origin, derived from rock minerals, or chemically synthesized. Primitive agricultural methods quickly exhausted the soil's original content of one or more of these essentials, and croplands were commonly abandoned after a few years in favor of new localities. As the science of agronomy began to develop, better methods involved replenishment of these elements by crop rotation or fertilization. Nitrogen is the critical element here. The other three of the four elements mentioned are available in practically limitless amounts in the sedimentary rocks of the earth's crust—the phosphates, limestones, and the evaporites, rich in potassium chloride. Although there is an inexhaustible supply of nitrogen in the earth's atmosphere, highly organized plants cannot utilize it directly from that source. The bacteria that grow abundantly on the roots of leguminous plants, however, can remove it

from the air and transform it into food for plants. Thus it
has long been the practice in scientific agriculture to grow
a crop of cowpeas, soybeans, or alfalfa and to plow it under
to restore fertility for a subsequent crop of cotton, corn,
or wheat. Replenishment of nitrogen is in reality the basis
for crop rotation. Nitrogenous nourishment for plants is
also supplied by animal excrement, the manure constitut-
ing the fertilizer first used by farmers of old.

Extensive use has also been made of the natural nitrates,
chief of which is the mixture of sodium and potassium
nitrate known as saltpeter. Prior to World War I, Chile was
the major source of this mineral, used both for fertilizer
and for the manufacture of gunpowder and other explosives. Deposits of "chile niter" are extensive but by no
means inexhaustible. During that war, Germany and her
allies were effectively cut off from all sources of abundant
supply of nitrogen except the atmosphere. It was imperative to maintain the flow of ammunition to their army,
navy, and air force. Under those conditions, German
chemists hastened to perfect a method of nitrogen fixation
whereby atmospheric nitrogen is converted into chemical
compounds useful not only in the production of high explosives but also in the synthesis of nitrogenous fertilizers.
Since World War I, facilities have been installed in many
countries throughout the world, most of them in fairly
close proximity to sources of hydroelectricity where the
necessary power for operation of the process is available
at relatively low cost. The bottleneck has been opened
wide; nitrogenous fertilizers are now a renewable resource. The products of the soil are a part of man's annual
income.

This means that the earth's croplands are to be considered as a factory that may be kept in operation throughout
an indefinitely long future rather than as a mine that will
be exhausted sooner or later. Therefore, the question is
whether the factory is, or can be made, large enough to

meet man's present and future needs for sustenance. Those competent to answer that question are in general agreement that if all the available good agricultural lands of our planet were worked by the best modern methods, the resulting supply of food would be more than adequate to sustain the present world population.

Efficient operation of the earth's food factory requires much knowledge about nature and considerable skill. During the last few decades, agronomists, agrobiologists, geneticists, and other highly specialized scientists have learned much from their research, and much of that knowledge has found practical application by the actual tillers of the soil in many regions. This is an important part of the cultural evolution of mankind; the future of that kind of evolution will be much longer than its past. Each new advance both in research and in practice opens doors for further progress, and no end to the road is in sight. Already soil productivity has been vastly increased in every region in which the inhabitants have been privileged to take advantage of the new knowledge and techniques. Consequently, during the late 1960s and early 1970s, the world production of such important food-stuffs as wheat, rice, and corn increased at a somewhat more rapid rate than did the world population. More food is now available per capita, the world around, than ever before. One small step has been taken on the long road leading to the ultimate elimination of human malnutrition and starvation from the face of the earth.

To be well nourished a human being requires a balanced diet of a variety of foods, not all of which can be produced from any given type of soil in any local climatic zone. Basic for that diet is an appropriate balance between carbohydrates and proteins. Food produced directly from cultivated croplands consists predominantly of carbohydrates; food harvested from seas and lakes is predominantly protein. The fisherman and the farmer are just as inescapably interdependent as are the air-traffic controller and the computer programmer.

The waters of the earth are as truly a part of the planetary food factory as are its lands. Knowledge and skill are similarly required for efficient operation. The biologic phases of oceanography are still in their infancy and the scientific technology of farming the sea lags far behind that of farming the land, but there is good reason for being optimistic about its future. Some of the bio-oceanographers are suggesting that the renewable food resources of the sea are adequate to meet the protein requirements of forty billion human beings! As a matter of fact, the total annual catch of the world's fisheries has been increasing in recent years at a rate significantly greater than that of the increase in world population.

All of this adds up to the conclusion that at the present time the real problems are not those of food production but those of food distribution. That conclusion is highlighted by the fact that the United States government has recently been paying American farmers hundreds of millions of dollars per year to keep their fertile acres fallow, even though hundreds of millions of persons, some of them in the United States, are undernourished or starving. Photographs of starving children and weeping mothers in various Asiatic and African countries, widely circulated in newspapers, magazines, organizational appeals, and television programs, drive home the desirability, if not the necessity, for prompt solution of those problems.

Equitable distribution of food among all mankind is definitely a human responsibility. Attainment of that goal within the framework of the terrestrial environment is entirely possible, given the determination to put aside political and racial prejudices. Any rational planning toward that end must take into consideration the fact that, regardless of the degree of perfection attained by food technologists, local and temporary famines will continue to menace

Bangladesh and Nigeria and crops will occasionally fail in America's midwest and the Soviet's Ukraine because of the vagaries of weather and climate. Some sort of world food bank should be established, presumably under United Nations auspices, in which the surplus from years of plenty might be stored for distribution to the hungry in years of scarcity. The inauguration in 1974 of a World Food Council, to have its headquarters in Rome, Italy, was a significant step toward effective international cooperation in the effort to increase the world's food supplies and distribute them more equitably.

There is of course an ultimate limit to the number of human beings whose needs for food can be met by the earth's renewable resources. The earth has a diameter of only 8000 miles, a surface area of only 197 million square miles; seven tenths of its surface is covered by water, only three tenths is land; and of the land somewhat less than half is even potentially suitable for the growing of crops. Nevertheless there remain many thousands of square miles of arid land that could be irrigated and marshlands that could be drained.

Now that practical techniques for desalinating salt water are available, the previous limitations on supplies of fresh water have become flexible. It is impossible to make even an educated guess concerning what might be the ultimate limit of the output of the earth's food factory. Two conclusions may, however, be drawn with confidence. Mankind today is nowhere near the limits set by nature for the physical well being of the human inhabitants of the earth. There is no reason for despair concerning man's technological ability to gain abundant sustenance from earth's renewable resources for a much larger population than that of today.

But the technologies of human sustenance will require, at least throughout the near future, increasing expenditures of nature's stored capital, the nonrenewable resources of the earth. Plows, tractors, cultivators, harvesters, and other kinds of farm machinery essential to the expansion and more efficient operation of the food factory are made largely of iron and other metals. So also are the trucks, freight cars, and ships used to transport the farm products to markets or storage facilities. Even the production of nitrogenous fertilizer from the inexhaustible nitrogen in the atmosphere requires machinery made at least in part of metals. Incidentally, the world's production of nitrogenous fertilizer increased from nineteen million metric tons in 1965 to nearly thirty-eight million metric tons in 1973. At present and for decades to come, increasing the food supplies from renewable sources inevitably hastens the exhaustion of the nonrenewable goods in Mother Earth's storehouse as the world population increases.

Looking toward the far-distant future, it is evident that mankind must eventually learn how to live within the annual income from renewable resources. How that may be done, no one can now foresee. The development of synthetic plastics manufactured from plants and/or lowly forms of animal life, rather than from petroleum or coal, and current research concerning artificial photosynthesis may be clues suggesting possibilities, but essentially it will involve presently unpredictable breakthroughs achieved by the science and technology of future centuries. The meaningful question for today is whether or not the available nonrenewable resources will prove adequate for human needs throughout the next century or two, a time which I hope may be sufficient for the achievement of those breakthroughs.

When I was talking and writing[1] about such matters in the 1940s, I affirmed that there was "enough and to spare" on the shelves of Mother Earth's cupboard to undergird human welfare for at least several decades, if not for a few centuries. I used the statistical data available at that time: statistics showing the recent annual consumption, for the world as a whole and for the most industrialized nations, of the significant metals and mineral fuels, statistics showing the proved and highly probable reserves of those items as known or estimated at that time, statistics showing the recent changes in population and the projections made by demographers concerning future changes. Looking back over the quarter century from the vantage point of the present time, it is evident that many of those statistics were in error. Populations have increased much more rapidly than was expected; per capita consumption of several significant items is greater than anticipated. More goods are being withdrawn from the storehouse each year than was then projected. On the other hand, more reserves have been made available, more of the concealed packages on the back shelves have been opened than was foreseen in the 1940s. The two kinds of error made in 1944 offset each other almost precisely; the thesis of enough and to spare remains true today.

It must be remembered that estimates of proved and highly probable reserves of mineral wealth of any kind must always be made in the light of currently used methods of exploration and exploitation. In the last quarter-century, newly invented tools and techniques, mainly geophysical in nature, have enabled geologists to discover oil pools containing amazingly large quantities of petroleum and natural gas in submerged portions of the continents off the Gulf Coast of Louisiana and Texas, in the North Sea and Cook Inlet, Alaska, off the shore of California and elsewhere, as well as in the Middle East and on Alaska's North Slope. At the same time, new and improved methods of drilling for oil, such as from floating barges or from multi-well platforms supported by tall stilts in fairly deep water, have combined with more efficient techniques of secondary recovery to increase greatly the earth's available mineral fuel resources. The ultimate exhaustion of conventional mineral fuels, both in the United States and in the world as a whole, will be postponed far beyond the dates I suggested in 1944.

In the present stage of mankind's cultural evolution, the need for abundant supplies of mechanical and electrical energy is at least as crucial as the need for food. Given a continuing supply of that kind of energy, the chemists can continue to work miracles with their molecules, the food producers can continue to improve the efficiency and increase the dimensions of the food factory, the physicians and public health experts can continue to enhance the physical welfare of human beings everywhere, the transition from major reliance on nonrenewable resources to major reliance on renewable resources can be effected. Insofar as energy sources are concerned, that transition is well under way at the present time. It is in fact being expedited in the mid-1970s by the attempt of oil-rich Arab states to use oil as a political bludgeon and by the abrupt increases in the price of crude oil decreed by the cartel known as OPEC (the Organization of Petroleum Exporting Countries). The dark cloud of an energy crunch that looms so large over the highly industrialized nations of the world has indeed a silver lining. Lack of adequate sources of mechanical and electrical energy will probably never be a roadblock across the beckoning highway of mankind's future.

Replacement of the mineral fuels by waterpower, geothermal energy, solar radiation, and nuclear fission as the major source of mechanical and electrical energy will, however, continue if not increase the demand for such metals as iron, copper, aluminum, lead, zinc, and tin. Are the nonrenewable and limited resources of metallic ore bodies adequate for the demands that will be placed upon them in the foreseeable future? In answering that question, it should be noted that at no time do the mining engineers ordinarily have in sight more than twenty or thirty years' supply of the metals they are exploiting. Geologists do well to keep that far ahead of demands. Estimation of highly probable reserves as distinct from proved reserves is inherently more difficult than for the mineral fuels and therefore less reliable. Again, many unexpected discoveries of new ore bodies, some of them of great magnitude, have been made since 1940 as a result of the application of recently developed methods of geophysical exploration. Among them are the rich iron ores of northeastern Quebec and southeastern Venezuela and the extension of Africa's copper belt in Zambia. At the same time, metallurgical research has made it economically possible to winnow the metallic content from many lower-grade ores that were formerly without value. The prime example is the taconite ore of the Lake Superior region. Formerly just a rock, useless except perchance as road metal, it is now a valuable iron ore that increases many fold the iron reserves within the United States and Canada.

The current situation with regard to production, consumption, and reserves of the metals and mineral fuels throughout the world has been surveyed recently by Charles F. Park Jr.[2] in an authoritative and illuminating book aptly entitled *Affluence in Jeopardy.* The statistical data Dr. Park presents are thoroughly reliable. Some of his estimates of highly probable reserves may be a bit on the conservative side, but that may well be the wiser way to treat them if readers are to be alerted to the poignant seriousness of the impending crisis. I cannot accept, however, his unqualified assertion "that poverty and starvation are bound to increase in the world, not because of its inability to feed more people [he does not discuss that problem] but because of potential shortages of many mineral products that are necessary for an industrial civilization and for modern standards of living." I wish he had added: "unless men do what they could do to prevent those potential shortages from becoming realities." Actually he proceeds to offer some cogent suggestions about what might be done in that regard.

The figures used by Dr. Park accurately indicate the tremendous increases in annual production of iron and steel, of copper, and of lead that would be necessary should all the world achieve the present per capita consumption of those metals in the United States. They are not, however, as alarming as they might appear to be at first glance. Such an achievement would inevitably be accompanied by a comparable increase in the number of competent geologists and geophysicists and of skilled metallurgists and mining engineers. The concealed packages on the back shelves of Mother Earth's cupboard would be opened at an equally faster rate. This would mean that the irreplaceable goods in nature's storehouse would be exhausted sooner rather than later and that the shift to greater reliance on the earth's renewable resources would have to be made within a few decades rather than within a few centuries. It is just possible that the breakthroughs required for that adjustment to the limitations of our environment might be made by research scientists of nations that are at present far behind in the trend toward modern standards of living. Perusal of *Pre-Investment News*, published monthly by the United Nations Development Programme, is especially reassuring. Thanks to international cooperation, formerly unknown or unused resources, some in the renewable category, others in the nonrenewable, are now being discovered and exploited at a truly gratifying rate. At the same time thousands of citizens of the less developed countries are receiving technical training and advanced education in their native lands to provide the skilled manpower necessary for those tasks.

Mankind has responded notably to the biblical directive to "be fruitful and multiply and fill the earth" (or "replenish" it and/or "subdue" it). Today there are scarcely any good lands left to fill. Overpopulation is a very real danger for the immediate future, if indeed human welfare is not already seriously endangered in certain over-crowded regions. So much has been said and written of late concerning this matter that I need not document it here. The number indicating the optimum human population of a planet having the physical characteristics of the earth is a matter of opinion among demographers. It ranges all the way from half the present population to twice or even thrice that number. The higher figures, however, are suggested more as possible maximal populations rather than as optimal; almost certainly they would involve something far less than affluence for most if not all people.

Now that the moon is distant from the earth by only three-and-a-half days of travel time, it has been suggested that earth's overpopulation might be alleviated by colonizing the moon, even as Europe's overpopulation in the nineteenth century was alleviated by colonizing North and South America. The idea is utterly ridiculous. I fully expect that by the end of the twentieth century, astronomical observatories will have been constructed on the moon's surface and manned the year around. There is even a remote possibility that mineral resources of such a unique nature or of such a quality as to justify the excessive costs of exploitation may be found on the moon. If so, mining camps may some day be established there. But the life-support systems for every person on the moon will necessarily be transported and serviced from the earth. *Homo sapiens* cannot adjust its anatomy to existence in any environment other than that of the planet which has mothered it. With all due deference to the ingenuity and efficiency of modern technology, the expenditure of energy and materials required to colonize a million acres on the moon's surface, even in the Sea of Tranquility, would be many times greater than that required to colonize a million acres of the Sahara Desert here on earth.

Or if you are thinking about the potential mineral resources of the moon, let me call certain facts to your attention. Many of the most valuable ore bodies we are now exploiting are in what we geologists call the Precambrian basement complex. As the term implies, these are relatively ancient, largely igneous or much-metamorphosed rocks, concealed beneath the surface rocks of every continent and exposed at the surface throughout some small fraction of the area of each. In North America the largest area of their surface exposure is the Canadian Shield. Throughout a still larger area of that continent they are covered by a veneer of younger, largely sedimentary rocks ranging up to ten or fifteen miles in thickness. Throughout much of the broad Mississippi Valley the basement complex is less than five miles below the present surface. It has been reached by thousands of wells drilled in the search for oil or gas, but ordinarily the drilling is stopped as soon as the drill penetrates to it; petroleum geologists know that rich oil pools occur only in sedimentary rocks. There is, however, no reason to believe that ore bodies such as the iron ores of the Lake Superior region or Quebec, the cobalt and nickel ores of Ontario, the copper ores of northern Michigan, or the uranium ores near Great Bear Lake in northwestern Canada are limited in their distribution throughout the Precambrian basement complex of North America to that portion of it which is exposed at the surface in the Canadian Shield. Exploration for mineral wealth a couple miles or more beneath our feet, and the exploitation of the basement complex there concealed, would be far less costly than exploration and exploitation of hypothetical ore bodies two hundred and forty thousand miles above our heads.

The conclusion is inescapable. Mankind's destiny is that of an earth-bound creature. Salvation must be sought here, on this terrestrial planet. To gain continuing existence, man must adjust himself more adequately to the environment in which he has long been living. That adjustment will involve two basic changes in his way of life. On the one hand, the recent rapid increase in population must be slowed down. Mankind must accept a new directive, now that the old one has been fulfilled. Quality of off-spring rather than quantity must be the aim. On the other hand, the resources of the earth must be conserved more wisely than they have been in the past. Wastage of nonrenewable resources must be reduced to a minimum; renewable resources must be used as intelligently as human capabilities will permit. It is far easier to announce these new directives than to live in accordance with them, but the current trends of thought and action encourage the hope that adequate adjustment to the sum total of environmental factors will be made.

Never before in human history has so much attention been given to the population problem as it is receiving today. More research[3] concerning the biology of reproduction and the techniques of contraception was under way during the 1960s than in any previous decade, and that research is continuing during the 1970s. The ideal contraceptive has probably not yet been discovered, and it may never be, but it is already possible for those who have access to the knowledge and materials of up-to-date biological science to undertake family planning with great confidence. One national government after another is committing itself to policies directed toward restraining population growth, and many others are almost certain to do so in the next few years. The United Nations' World Health Organization has accepted the idea that birth control must be of primary concern in its objectives for human welfare. It is certain that the percentage of unwanted babies born each year in coming decades will be less than in those that have passed.

All this, however, may prove to be just another instance of too little and too late. As I see it now, the odds are about even as to whether the population problem will be solved in time to save mankind from catastrophe. The less-than-satisfactory outlook toward the future drives home the fact that all human beings are in this predicament together; *each of us is a member of the family of mankind.* The key words in the preceding paragraph were those pertaining to access to the knowledge and materials of modern biology. Sharing the benefits of science and technology among all the inhabitants of the earth is prerequisite to any shifting of the odds of favor of mankind's future.

The second of the new directives pertains to the ways in which men and women use the earth's resources. The record of the last few centuries has not been very good, even though we may point with pride to the greater efficiency and increased comfort made possible by science and technology. Too often and in too many places, those resources have been used selfishly for the satisfaction of an individual, a corporation, or a political unit without regard for the consequences to other people or to future generations. The delicate ecological balances of nature have been heedlessly and unnecessarily disturbed by activities that at the time and place seemed praiseworthy. Irreplaceable metallic ores have been expended for destructive rather than constructive purposes. Air and water have been polluted by noxious fumes and harmful waste for which men are responsible. Urban sprawl is ruining many a formerly enchanting landscape. One is tempted to say that mankind has been despoiling rather than replenishing the earth.

Fortunately, the concept of conservation of natural resources is today spreading widely throughout the world and deepening its hold upon the minds and hearts of people everywhere. Although in its etymology and early usage, conservation implies preservation and protection of something, the conservation of natural resources, as the term is employed today, puts the greater emphasis upon their use. Renewable resources should be used by the present generation in such ways that they will also be available for future generations. Nonrenewable resources should be used for the welfare of mankind as an all-inclusive and on-going entity, even though they will ultimately be exhausted. Thus the preservation of the status quo is only a minor element in the broad concept of conservation. The scenic beauty, geological features, and wild life of national parks or monuments, state or municipal parks, and game preserves or sanctuaries are to be protected in order that they may be useful to oncoming generations. Usefulness here is appraised not in the quantitative terms of material benefits or economic gains, but in the qualitative terms of intellectual outreach, esthetic awareness, or ethical consciousness. Rarely does anyone grow in knowledge of nature unless he has contemplative contacts with nature unspoiled by human hands. In a similar vein, the modern city planners include in their blueprints such open spaces as the greenbelts peripheral to the inner core of cities like London or the commons inherited from colonial days by many New England towns. These are useful not only for the health of human bodies but for the health of mind and spirit as well.

Pollution of the air, water, and soil by automobile exhausts, effluence from factories and smelters, and waste products from biological activities are of serious concern to conservationists, affecting as they do the earth's renewable resources. So also are pesticides and other chemical substances that start chain reactions among the ecological balances of nature, some of which may result in unexpected and disastrous consequences. It has become abundantly apparent that technological innovation is a decidedly mixed blessing. Degradation of the human environment by pollution and ecological disturbances, combined with over-crowded cities, jammed highways, and spoilation of landscapes poses a danger for the future of mankind at least as ominous as a nuclear cataclysm or the exhaustion of irreplaceable mineral resources. These are just as much a consequence of the application of scientific knowledge in industrial technology as are the marvelously increased efficiency and improved living conditions of modern machine-using man. It would, however, be as inaccurate to think of modern technology as the scapegoat of all social ills as to take the euphoric view that technology is competent to guarantee universal felicity. What is needed is greater wisdom in making prior evaluations of the adverse, along with the beneficial, effects of contemplated innovations, an art now known as technologic assessment.[4]

Such wisdom is not easy to come by. It requires both superior intelligence and a keen desire to contribute to human welfare—not just the welfare of an individual, a group, or a nation, but of all mankind; not only of the present generation, but of future generations as well. Thus far, in the United States and many other countries, technologic assessments have been made by competing industrial corporations, departments of defense, and other governmental bureaus, largely in terms of supply and demand in the economic market and on the basis of cost-benefit studies with limited perspectives. To take the broader, longer-range view, and to act in accordance with the directives it reveals is an expensive business. The costs of strip-mining for coal, for example, are greatly increased if the ridges of broken rock and disrupted soil left by the earth-moving machinery after the coal has been removed are smoothed over, covered with good soil, and planted with grass or saplings or otherwise prepared for the cultivation of crops. As things are today, that extra expenditure by mining companies in the competitive market will be undertaken only if there are governmental regulations to enforce it or governmental subsidies to underwrite it.

In general, when choices of alternative technologies must be made among competing and conflicting interests and values, it will often be necessary to make and enforce the selection collectively rather than individually. Politicians therefore become at least as responsible as scientists and technologists for implementing the results of technologic assessment. Even though they sometimes respond to it reluctantly, politicians are very sensitive to public opinion, even in an autocratic society. In the last analysis, therefore, we must depend upon enlightened public opinion to screen out the deleterious effects of technologic innovations and assure their beneficial consequences. The problems are complicated, entangled, and obdurate, but not insolvable. A glance at such periodicals as the *Pollution Equipment News* indicates that significant steps have already been taken toward meeting the challenge. Whether motivated by genuine altruism or enlightened self-interest, many technologists, industrial executives, and governmental administrators are now devoting thought, energy, and money to this important aspect of conservation. It will not be an impassable roadblock to mankind's future.

Turning now to the nonrenewable resources, it should be noted again that conservation means the wisest possible use rather than vigilant preservation. Debatable questions will inevitably arise as to priorities for allocation of the metals extracted from ore bodies in terms of wisest use for human welfare, but the asking of those questions is itself a mark of progress. And there will surely be widespread agreement to at least a few of the answers. For example, the leadership of the United States among the nations in per capita consumption of iron and steel (about one ton per year) is due in considerable part to the extensive use of that metal in the weaponry and other material of the U.S. Department of Defense. How much better it would be, simply in terms of conservation of irreplaceable resources, if the world could be so organized under international law that a world police force would make unnecessary any national armies! That of course is at present an unrealizable dream, but any reduction in armed forces resulting from international agreement for limitation of armaments would be a forward step in conservation of resources quite apart from ethical or moral considerations.

Conservation of natural resources is inherently future-oriented; it necessarily involves planning. And planning is a fine art as well as a recondite science in which man has not in general reached any great heights of excellence, even though the recent success of manned flights to the moon and back suggests that he is capable of doing so. Some thirty years ago, Karl Mannheim, an eminent German sociologist who spent his later years on the faculty of the London School of Economics and Political Science, described the history of thought as comprising three stages: the stage of chance discovery, the stage of inventing, and the stage of planning.[5] In the stage of chance discovery, he wrote, "some individual or group discovers accidentally, among a very large number of possibilities, the kinds of reaction which fit a given situation. The achievement of thought then lies in remembering the correct solution which has been discovered." The social life of the primitive food-gatherers and hunters was a result of this kind of thinking. No "precise, reflecting knowledge of the environment" was required. "Even today we react to many situations with a type of thought and conduct which is still at the level of 'chance discovery.' " When however "single tools and institutions were consciously modified and then directed toward particular goals," man's thinking advanced into the stage of inventing. "At this level man had to imagine a definite goal and then think out in advance how to distribute his activities in a given way over a certain period of time with this goal in view." It was not necessary to think beyond the task immediately at hand, but he had to imagine how his invention would fit into the immediate environment and "foresee the most probable consequences of an event." The development of technology has been largely within the framework of this kind of thinking. At the present time, man and society are gradually advancing "from deliberate invention of single objects or institutions to the deliberate regulation and intelligent mastery of the relationship between these objects." In this dawning stage of planning, man must not only give thought to particular aims and immediate goals but must also consider the "effects these individual aims will in the long run have on wider goals. The planned approach does not confine itself only to making a machine or organizing an army but seeks at the same time to imagine the most important changes which both can bring about in the whole social process."

Mannheim implies that the ascent from the level of inventive thinking to planned thinking is an essential part of the cultural evolution of man and that mankind will continue in the stage of planning throughout the foreseeable future. With this I heartily agree, but as I near the end of this chapter I want to present some thoughts of my own for which he should not be held responsible in any way.

The requirement for success in the stage of discovery is a good memory; in the stage of invention, a vivid imagination. Both of these intellectual characteristics are needed in the stage of planning. In addition there must be profound understanding, based on precise knowledge about the universe and man, as well as deep compassion that leads a man to say "I am not so much my brother's keeper as I am my brother's brother." With such requirements it is not surprising that would-be planners have made mistakes, or at the least have not fulfilled all expectations. The stage of planning is at the dawn. We should no more expect that the pioneer planners could come up with a perfect and workable blueprint of a new world of law and order, bringing peace with justice for all mankind, any more than the Wright brothers should have been expected to design a supersonic jet-propelled airplane at the dawn of aviation. Progress in planning will be by the trial and error method just as it has always been in the process of organic evolution. Most importantly, in the dawning age of planning we must plan for freedom as well as for orderly and equable distribution of the benefits of science and technology.[6]

Individual freedom means personal responsibility. The earth's resources are of such a nature and are available in such quantities as to permit man to look forward to the possibility of an indefinitely long future of comfortable and secure existence for himself and his kind. But the opportunity now afforded him will not be realized unless he accepts responsibility for his collective well being. Conservation of resources in order that the future of mankind may be long and bright will be accomplished only if enough individual men and women adopt the mental and emotional attitude of stewards of the earth's bounty. The sense of personal responsibility and the ideal of stewardship are attributes of the human spirit; to develop and strengthen them, men of religion should join hands with men of science as co-workers for the future of man as an inhabitant of the earth.

NOTES

1. Kirtley F. Mather, "The Future of Man as an Inhabitant of the Earth," *Scientific Monthly* 50 (1940): 193–203; also, *Smithsonian Institution Annual Report for 1940*, 215–29. *Enough and to Spare* (New York: Harper, 1944). "Petroleum: Today and Tomorrow," *Science* 106: 603–9; also, *The Advancement of Science* 4 (1948): 292–300.

2. Charles F. Park, Jr., *Affluence in Jeopardy* (Freeman, Cooper and Co., 1958).

3. Cf. Roy O. Greep, editor, *Human Fertility and Population Problems* (Schenkman, 1963).

4. *Technology: Processes of Assessment and Choice*, The Report of a Committee of the National Academy of Sciences on Science

and Public Policy, Harvey Brooks, Chairman (Washington, D.C.: 1969).

5. Karl Mannheim, *Man and Society in an Age of Reconstruction* (New York: Harcourt Brace, 1940) 150–55.

6. For Mannheim's views on freedom and equable distribution see Karl Mannheim, *Man and Society. . . .*

Religion in
the Permissive
Universe

DICTIONARIES TELL US that the word, religion, is probably derived from the Latin *religare,* to tie back, tie up, tie fast. Certainly in its earlier usage, the word implied something that would tie men and women together in brotherly love and noble aspirations and would unite individual men and women with the administration of the universe in respectful adoration and dutiful obligation. Thus the Christian joins with Jesus in praying (John 17:21) "that they all may be one" and with Paul (II Corinthians 5:19) in saying "that God was in Christ, reconciling the world unto himself. . . ."

But it hasn't worked out that way. Throughout many centuries of human history, religion has appeared to be much more a divisive factor than a unifying force in social relations. Christians and Muslims butchered each other during the crusades of the Middle Ages. The massacre of the Huguenots on St. Bartholomew's Day in 1572 was but one episode in the long series of religious wars that disrupted Europe in the sixteenth century. The barbarous tactics of the Spanish Inquisition were directed primarily against Judaism and Protestantism in the seventeenth century. Even into the twentieth century many an international conflict was viewed, at least in the eyes of some of its belligerents, as a Holy War. The cleavage between Northern Ireland and the Irish Free State is due in large part to religious animosities between Catholics and Protestants. The separation of Pakistan from India is the result of enmity between Muslims and Hindus. Semitic Muslims and semitic Jews continue to fight each other in the Middle East.

Fortunately the tide has turned and during recent decades Protestants, Catholics, and Jews are working together amicably and constructively in many communities and a few countries throughout the world. City, state, national, and international councils of churches, largely but not exclusively of the Protestant faiths, are serving effectively to unite churchmen of many previously competing if not antagonistic denominations in their commendable social and charitable activities. The ecumenical movement is growing in strength and influence throughout all Christendom and is beginning to appear among Buddhists as well. How much this new spirit of cooperation in religious affairs owes to the scientific enterprise is a matter of opinion, but in my judgment the debt is considerable. The organization of international congresses and unions of scientists began in the nineteenth century (the International Geological Congress, founded in 1878, was one of the first), and today there is scarcely a scientific discipline in which its devotees lack an international association. The United Nations Educational, Scientific, and Cultural Organization (UNESCO) was brought into existence without benefit of clergy and, among other things, provides an umbrella under which many of these international scientific bodies are sheltered. Men of religion whose concern includes the concept of mankind-united have not been oblivious to the unifying and fertile associations of men of science without regard to color, nationality, or economic circumstances. There is still hope that religion will actually make its appropriate contribution to the binding together of mankind in amity and aspiration.

The record of religion as a factor in reconciling man and the administration of the universe is similarly spotty. All too often, religionists have sought escape from the harsh realities of the world of sense perception to some other, usually supernatural, world in which there are no problems. When a Christian sings "I'm but a stranger here, heaven is my home" and uses those words in the same literal sense as they were used by the author of that song, he has abandoned the attempt to bind man together with the administration of the universe, responsible as it is for the physical as well as the spiritual aspects of nature. The same is true of the Hindu who stoically contemplates nirvana as he maintains the lotus posture beneath a banyan tree while flies crawl in and out of the eyes of small children by his side. By and large, science has probably done more than religion to make men feel at home in the universe and to reconcile them with its administration.

Fortunately, there is today a definite trend among leading clergy and laity in many Christian denominations, and among many Buddhists and Hindus as well, toward the replacement of the former escapism by a healthy acceptance of the world as it is. During the last fifty years the social gospel has had a remarkably widening spread and deepening hold among religionists in many denominations. The early slogan to the effect that "it is not enough to pluck a brand from the burning, we must put out the fire" is no longer heard, but the idea underlying those words is firmly established. Science has become the handmaiden of religion not only to increase the efficiency of the charitable enterprises it engenders but also to increase the credibility of the valid spiritual concepts it should uphold. In this age of science, religion has a greater opportunity than ever before to exert a beneficial influence upon the cultural evolution of mankind.

The evolution of religion and the development of science have been closely intertwined throughout all human history. Indeed, the academicians of the middle ages considered theology to be the queen of the sciences. To gain and hold a considerable number of committed devotees, any religious doctrine must seem rational in the context of the world view or natural philosophy of the time and place. It is a far cry from the anthropomorphic tribal gods of ancient Mesopotamia and classical Greece and Rome, with their sacrificial ceremonials, to the one universal god, whether Jehovah or Allah, of later times. But it is just as far a cry from the tiny vestpocket geocentric universe of ancient times, with its flat-earth-disk beneath an overarching firmament across which moved the sun, moon, and stars, to the vast heliocentric solar system of Copernicus, Galileo, and the eighteenth century astronomers. The clock-like universe of nineteenth century theologians was a reasonable extension of the mechanistic materialism of the contemporary scientists. The description of God as "the ground of being" and of religion as that which deals with man's ultimate concern is quite in keeping with the twentieth century discovery in science of the presence of nonmaterial realities in the human environment. To denigrate religion because it had its roots in tribal taboos and its early practitioners were witch-doctors and magicians is as unjustified as to cast aspersions upon astronomy because it stemmed in part from Egyptian astrology, or upon chemistry because it grew out of medieval alchemy.

The forward movement of both religion and science has been cyclical. In religion, it is the cycle of prophet and priest recurring again and again: the prophetic voice proclaiming a new vision and a new or revised doctrine, followed by the priestly cohorts who embalm the vision in icons and perpetuate the doctrine as dogma. In science, it is the cycle of innovator and traditionalist, likewise recurring again and again. The noun *atom* has had as many sequential meanings as the noun *god*. One reason for the continuing tension between religion and science is the willingness of post-Aristotelian scientists to break away from traditional doctrines (they prefer to call them theories or principles and shun the idea of dogmas) in contrast to the vested interest of religionists in the priestly rituals and codes. We often use traditional scientific language although we are fully aware that the words have new meanings which sometimes are in complete opposition to the earlier ones. For example, we still refer to sunset even though we know that the sun disappears at eventide because the horizon rises, not because the sun moves downward. Yet many of us fear that we will be charged with heresy if we use the symbolic language of our religious establishment in unorthodox ways.

What then is the place accorded to religion in relation to the world view of modern science? And what role may be assigned to it in the permissive universe? I want to speak first of religion in general and second of the particular religion that we designate as Christian.

The key words of religion in general are *God, soul, sin, virtue, salvation,* and *destiny.* That rules out the communist ideology of Marx and Lenin, even though many of its adherents are religiously devoted to it, as well as the religion of humanism which I consider to be secular philosophy; in neither is *God* a key word. As a definition of *God,* with a capital G, in the context of what I have said about the nature and administration of the world in which we live, I would suggest the following: *God* is a symbolic term used to designate those aspects of the administration of the universe that affect the spiritual life and well being of mankind.

I do not object to the use of such capitalized pronouns as *He* or *Thou* or *You,* or even the capitalized noun *Administrator* in referring to God, provided it is understood that such a reference is justified not by scientific knowledge about the administration of the universe but by spiritual experiences that defy quantitative measurement. The significance of the personal pronouns may run a broad gamut from something similar to that implied when the mariner speaks of his yacht as "she" to the much more profound meaning conveyed by the prophet who reports that he had "wrestled with God."

Returning to my definition of God, after that brief but important detour, He is thus a creative and regulatory power operating within the natural order about which we now have abundant, but by no means complete, knowledge. He is spirit, not attenuated matter nor an idealized man-like figure. He is immanent, permeating all of nature, unrestricted by space or time. He is transcendent only in that His spirit transcends every human spirit, possibly the sum total of all human spirits melded together. He is not supernatural in the sense of dwelling above, apart from, or beyond nature. Knowledge *about* Him is gained by studies that are essentially historical in kind. Knowledge *of* Him is gained by experiences that are essentially personal, generally involving moments of insight that yield discoveries and revelations. As I have already indicated, these are two different appellations for what is inherently the same phenomenon.

I do not equate God with the spiritual force field postulated in chapter 6, for two good reasons. In the first place, religion should never confine itself within a prison constructed of scientific principles or doctrines no matter how widely and firmly they may be held by men of learning at any time or place. The confrontation between the inquisitors and Galileo, or between Bishop Wilberforce and Thomas Huxley, should be a sufficient lesson for twentieth century theologians. The world view of modern scientists, with its recognition of the reality of the nonmaterial, provides much more fertile ground for the cultivation of religious concepts than did the mechanistic materialism of nineteenth century scientists; but those concepts cannot depend for their support on any scientific theories that permit their growth. Specifically in this case, nobody knows how long field theory will continue to be a valuable tool for research physicists. Already there are indications that certain details of that theory, at least, must be revised. Even more significantly, some of the theoretical physicists are now considering the possible reality of tachyons—astonishing entities that travel faster than the speed of light, accelerate as they lose energy, and may be everywhere at all times.[1] Concepts of nonmaterial realities will doubtless long endure in the physicists' repertoire, but the precise nature of the terminology and equations of those concepts a generation hence cannot be predicted. In the second place, I would not equate the aspects of the administration of the universe that affect the spiritual life and well being of mankind with the potential forces of the spiritual field alone, even as currently postulated. The directives in that field are best discovered or revealed by observational and introspectional study of human behavior, but the observational study of the behavior of other animals suggests that there also are foregleams of spiritual qualities in their lives. For example, the loyalty of a dog to its master, or the willingness of a baboon[2] to sacrifice its own life in order that other members of its troop may be protected from danger. Moreover, it may well be true that directives and regulations of other force fields also have something to do with the spiritual factors in human life. Certainly there can be no response to the directives of the spiritual field until a creature has become appropriately organized in response to the directives of the organic field so that it may become aware of the nonmaterial realities in its environment. It is best to leave my scientifically respectable definition of God in its generality without additional specifications.

Several terms commonly used in religious writings or

conversation as more or less synonymous with God convey ideas closely similar to that implicit in my definition. Thus, Infinite Spirit or Eternal Spirit suggest a nonmaterial entity, unlimited by space or time. Holy Spirit and Divine Spirit add the connotation of something worthy of awe and reverence. All of these are antithetical to human *spirit* or *soul*, the second of the key words I listed in the earlier paragraph.

Enough is now known about human nature to validate the concept that each human being is an indivisible unity composed of body, mind, and spirit. The red triangle of the Y.M.C.A., with those three words on its sides, was originally intended to be a symbol of man rather than of the organization which adopted it as its logograph. I know of no scientifically verifiable data that would support the idea that the human soul is a separate entity inserted from above or from without into the human body and residing therein during a person's lifetime. On the other hand, the evidence is compelling that the human spirit manifest during the life of an individual may have a powerful effect upon the lives of countless other human beings, unlimited by space or time, after the individual's death. Equating thus the human soul with the spiritual aspects of the life of man, it follows that the soul, like the body or the mind, is a product of evolutionary processes.[3]

As noted in chapter 7, there were lowly creatures, far back in our ancestral lineage, who were unaware of the presence in their environment of either the electromagnetic or the gravitational fields. It made no difference to their welfare in which direction there was light or darkness or which way was up or down or north or south. Or if it did, they were impotent to take advantage of it. Before the beginning of the Cambrian Period, however, multicellular animals had emerged in great variety. Most of them had developed organs, structures, or behavior patterns that responded to the regulations of those two fields. Presumably, awareness of their presence as a part of the environment preceded response to their directives. Anatomical responses to the electromagnetic field had obvious survival value. It is not surprising that discrete organs, such as the eyes of trilobites, had evolved as far back in time as the Cambrian Period. The human eye is an inheritance from unknown ancestors who first became aware of a small portion of the electromagnetic spectrum.

Much the same can be said about responses made by organisms to the gravitational field. A free-swimming streamlined creature, like a fish, able to keep itself right side up in its aquatic environment, has a much better chance of surviving than one that is completely unaware of the earth's gravity. Indeed, the lateral-line organs of Paleozoic fish seem to be the antecedents of the much more efficient vestibular apparatus in the inner ear of mankind, one of the functions of which has to do with the maintenance of equilibrium. Thus the athlete can walk on a taut wire or perform on a trampoline because he has inherited from a long line of ancestors the ability to respond to the earth's gravitational field.

In contrast, none of the creatures in man's phylogenetic lineage seems to have been aware of the earth's magnetic field. Presumably their survival was unaffected by their inability to know which way was north or east or any other geodetic direction. Today the Boy Scout carries a magnetic compass lest he lose himself in the forest. Man has compensated in characteristic fashion for this deficiency in his heritage by inventing the compass and devising radio beams in LORAN systems.

The evolution of the soul, or human spirit, in response to the presence of the postulated spiritual field may be compared to the evolution of human eyes and inner ears in response to the electromagnetic and gravitational fields and of the compass in response to the geomagnetic field. There should be no expectation, however, of finding any anatomical component of the human body, like optic nerves or vestibular structures, that might be the seat of the soul. The spiritual aspects of human nature are involved in man's cultural evolution, not in his biological evolution, except insofar as they depend upon his mental development which in turn depends upon the physical characteristics of his brain.[4] So also are the magnetic compass and the countless other artifacts that mark the various levels of attainment in man's cultural development. Indeed, an analogy may be drawn between the instruments and inventions of technology that increase and extend the efficiency of the human body on the one hand and the artistic creations and eleemosynary institutions that express and extend the nobility of the human spirit on the other hand. All of which is to say that the human soul is an achievement, not a gift.

Third in my list of key words of religion in general is *sin*. Failure to think, speak, or do what one ought to think, speak, or do is a sin of omission; thinking, speaking, or doing what one ought not to think, speak, or do is a sin of commission. The problem then becomes one of moral obligation, duty, or constraint. A first approximation to the solution of that problem may be found in the context of mankind's evolutionary development. *Homo sapiens* shares with all other species of animals the prime purpose of maintaining the continuous existence of its kind of life as long as possible. Individual or collective actions that deter mankind from achieving that objective are therefore sinful. This of course is a glittering generality but it can be brought somewhat into focus by asking the question it implies. In the present cosmopolitan phase of human evolution, with the potential abundance and inescapable interdependence of our technological culture, with the storm signals of overpopulation and genocidal wars on the horizon, and with the revolution of rising expectations already beating on our doors, what kind of interpersonal attitudes, intracommunity relations, and international policies are most likely to facilitate the orderly and equitable exploitation of the earth's rich resources for the welfare of all mankind throughout the foreseeable future? It is a long, hard, broadly involved question, but any rational well-informed person who honestly seeks its answers will certainly find among them important elements of the ethics he should live by. Failure to apply those ethical principles is sin.

But the evolutionary context reveals more than that. Throughout the last half-dozen millennia, if not indeed for a much longer time than that, *Homo sapiens* has been involved in the noble experiment of preserving and enhancing the unique values of individual members of the species within the framework of its collective purpose. Each of the world's great religions is committed to that endeavor, either explicitly or implicitly.[5] Essential to the success of that experiment is the continuing enrichment of the spiritual aspects of human life, the evolution of the soul, as a part of man's cultural evolution. The much-used cliché, "man does not live by bread alone," suggests that the flowering of the human spirit is prerequisite to the continuing existence of mankind. Certainly the evolutionary trend of the recent past may be accepted as indicative of the directives for the immediate future. Therefore thoughts, words, or deeds that blight or hinder the spiritual enrichment of the life of any fellowman are sinful.

I have chosen *virtue* rather than *righteousness* as the key word antithetical to *sin* because its overtones are more dynamic. Virtue is more than the absence of sin; it involves a commitment to positive thoughts, words, and actions that contribute to the spiritual development of others and thereby of oneself. It reaches deep into motives and intentions. The broad generalizations offered in the two preceding paragraphs as criteria for defining sin are exceedingly difficult to apply in practice. It is doubtful whether the most knowledgeable individual or committee can at any time foresee unerringly the ultimate consequences of contemplated action. Fortunately that is not required. The processes of evolution have always dealt solely with immediate and proximate consequences. It is so for man in his moral dilemmas. The virtuous man gives thought to the consequences of contemplated action as far ahead as he can see them; for the unforeseeable future he has confidence in the directives implicit in the spiritual aspects of the administration of the universe. With Henry Hudson, as imagined in Van Dyke's poem, he will "keep the honor of a certain aim amid the peril of uncertain ways and sail ahead, and leave the rest to God."[6] He must be sure, however, that his aim is in accord with our two criteria and that, in this time of rapidly increasing knowledge in the behavioral sciences, he is actually seeing as far ahead and as far around as he can.

The fifth of my key words of religion in general is *salvation*. In the context of human evolution, the salvation of *Homo sapiens* means the preservation of the hominid taxon from extinction. This is obviously permissible within the framework of all known administrative regulations as long as the earth continues to be a suitable abode for life as we know it. Success in achieving such continuity of existence depends in part upon the willingness and ability of human beings to adjust their lives to the changes in their environment that are taking place today and will take place in the future. Because most of those changes are of man's own making, such success depends also upon the wisdom man displays in making his environment nearer to his heart's desire. On both counts, salvation in the sense I am using it here is mankind's responsibility. To use the ecclesiastical terminology, this kind of salvation is "salvation by works." What might be in store for mankind when the earth is no longer fit to be the dwelling place of living organisms is a meaningless question for the man of science. And I would hope that the man of religion would not waste time worrying about it.

Far more important for religion is the salvation of human souls, the perfecting of the spiritual aspects of the lives of individual human beings. The spirit of man shares with his body and his mind both the handicaps and advantages of his heritage. Considering either man's biological or cultural evolution, that heritage is a mixed bag of odds and ends. It includes among its many items amity toward some and enmity toward others, suspicion of the new or strange and confidence in the old and well known, compassion for those in need and sympathetic consideration for the rights and needs of others as well as selfish greed or rugged individualism. Most important of all are the various natural tendencies toward or against the acceptance of personal responsibility for the welfare of one's companions and of mankind in general. Souls are saved when the human spirit is liberated from the degrading factors in its heritage and environment, so that it may grow in harmony with the directives of the spiritual force field. Salvation, in that sense, is often stimulated by charismatic speakers or writers, but central in the process is personal knowledge of spiritual aspects of the universe gained by open-minded contemplation and patient introspection.

The last of my key words is *destiny*. Biologists are fond of telling us that man is the only creature known to them to be capable of determining his own destiny. This of course is true only within limits. What they mean is that man has the capacity to select one of the several destinies available in the permissive universe and then to work his way toward it. To identify the most glorious of man's possible destinies and guide him toward its attainment is a function of religion. From the point of view of the student of geologic life development, man's noblest destiny is union of the human spirit with the spiritual aspects of the administration of the universe in the cooperative task of continuing the cultural evolution of mankind. As thus conceived, man's destiny is not a future condition of blissful rest but a dynamic state of continuing activity in which things of cosmic importance may be done. Most distinctive in that state is the unifying relationship between man and God, using that symbolic word in the sense that I have used it in a preceding paragraph.

To attain such a destiny, mankind must continue the trend of the recent past toward increasing awareness or consciousness of the nonmaterial factors in the cosmic environment. The potential range of human consciousness should be further investigated. Extrasensory perception is in large part amenable to rigid scientific research, and the parapsychologists have by no means reached the end of their road. More knowledge should be forthcoming from the biochemists and their colleagues in psychology concerning the effects of the hallucinogenic drugs that are claimed by some to be "consciousness expanding." Euphoric fantasies, however, do not provide the means for realistic increase of man's awareness of his cosmic environment. Moreover, as every sensible person knows, meddling with the chemistry of the brain and nervous system is fraught with grave dangers. Much more useful is the knowledge of human nature and of the forces operating in the world around us that may be gained by introspection and contemplation. Meditation is a fine art that requires training and discipline comparable to that required for successful research in any of the sciences. Many of the sages in the eastern civilizations, ancient and modern, were or are better at it than most of us in the western civilizations. It is, however, the only way to get really in tune with the Infinite, to become truly at one with the administration of the universe. The ancient advice is valid today: "Be still, and know that I am God."

In practice, the attainment of the kind of destiny considered in preceding paragraphs means whole-hearted allegiance to the concept of mankind[7] and thorough-going acceptance of the role of steward. The resources of the earth, including its human resources, are to be used for the welfare not only of the present generation but of all future generations. Thus, men and women, with their manifest creative powers, join the administration of the universe in projecting forward the trend of human evolution that characterizes the recent past.

The destiny of man as depicted in most of the world's religions involves more than this. At many times and in many places the prophets and priests have had much to say about the immortality of the human soul, and beliefs concerning that concept are deeply embedded in the minds and hearts of countless true believers today. What can a religion that is respectable in the light of the contemporary explosion of knowledge say about that idea? The question is so completely beyond the ken of the scientist *per se* that I approach it with great hesitation; I have, however, gone so far in dealing with arcane mysteries of life that I would be craven not to tackle this one.

Granted that the human spirit is an entity distinguishable from the human mind and body, several speculations concerning its viability are permissible. One of these encompasses the idea that the spirit is so dependent upon the mind and body that it ceases to function when the body dies and the neurons no longer activate the brain. The spirit manifest during any person's life may of course live on in the memory of those who knew him, in his literary or artistic productions, or in his influence upon political, religious, educational, industrial, recreational, or esthetic institutions and organizations. Shakespeare would doubtless agree that some of the good as well as "the evil that men do lives after them. . . ." In fact, the Immortal Bard is himself a prime example of this kind of immortality. It is something to which every person should give heed, especially in his moments of decision-making.

A quite different yet also permissible speculation starts from the assumption that the human spirit may be so nurtured and cultivated by the person possessing it that it develops an autonomy of its own, distinct from the body and the mind. The spiritual aspects of a person's life thus become his soul in the religious sense of that word. The speculation then goes on to suggest that when a person dies, the soul is freed from the bonds of space and time and may continue its existence in some other dimension or condition. Under the new and different regulations it loses it own identity and merges with other souls to form a universal spirit or spiritual continuum. This, I take it, is something akin to what Teilhard[8] has called "the noösphere," although I doubt that this speculation correctly denotes his concept of immortality.

A third permissible speculation starts from the same assumptions as the second but goes on to suggest that the soul retains its own identity after it is freed from the bonds of space and time. Many persons have reported from many times and places that they have seen such departed souls either as disembodied spirits or in human garb. Many have claimed that they have communicated with such inhabitants of the spirit world. But so far as I know, there are no reliable scientific data that support the credibility of those reports and claims. Many of them seem to be the result of hallucination or fantasy, or a product of the subconscious mind, as in day dreams. It may be that continued investigation by the members of the several societies for psychical research and their colleagues in parapsychology will yield trustworthy information before long concerning this area of human experience.[9] But in the meantime this concept of human destiny remains in the domain of religious faith. Insofar as the doctrines stemming from that faith avoid the materialistic trap set by geographic and climatologic concepts of hell, purgatory, paradise, and heaven, and the biologic pitfalls of reincarnation, they cannot be considered as unreasonable in the context of our present knowledge concerning the permissive universe.

I have selected Christianity to be critically considered as a specific variant of religion in general for a simple reason. The majority of the persons, young and old, with whom I have associated over the years have been Christians, at least in name. Moreover, it is the religion in which I was reared by devout parents; since my boyhood I have been a member, "in good standing" as far as I know, of a Baptist church affiliated with the American Baptist Convention and the National Council of Churches of Christ in the United States of America.

One might expect that Christianity would be the religion taught and practiced by Jesus of Nazareth, known to many as the Christ, who lived and died in Palestine nearly two thousand years ago. But it has by no means turned out that way. The dual process of misinterpreting his teachings and confining his ideas within more or less extraneous doctrines began even before his death. He authored no known documents; the only writing in which he is reported to have engaged consisted of symbols traced by his finger in the sand (John 8:6–8). Evidently he was so confident that his spirit would live on for countless generations in the hearts and minds of his followers that it was unnecessary, and perhaps unwise, for him to spell out a written testament, draft a covenant, or compile a code of behavior (Matthew 28:20). The earliest documents now available that might be useful to the historian concerned with what he actually said date back to just several years after his death. Most of these were written by men who like himself were devout adherents to the Jewish faith, and they took special pains to indicate that certain events took place and certain things were said in order "that the scripture should be fulfilled" (John 23:34, 36; *et al.*). To what extent their descriptions of events and recollections of his sayings were colored by their desire that this might be so we cannot know.

Especially significant in this context are the amendments

to Jesus' teachings and additions to his doctrines made by Paul (Sha'ul of Tarsus) in his letters now incorporated in the Christian Bible. Almost certainly, Christianity would not have become a powerful force in the Holy Roman Empire and a world religion of great renown had it not been for the self-sacrificing devotion and missionary zeal of this most influential "servant of Jesus Christ" (Romans 1:1). Nevertheless there are certain items among traditional Christian beliefs for which Paul is responsible but which do not seem to stem from the reported sayings of Jesus or to be attributable to him. Paul accepted the Garden of Eden account of creation in the Torah (Genesis 2:4–3:24) as a literally true historical record rather than as allegorical folklore.[10] Thence came, at least in part, the doctrine of the "fall of man" and the belief that man had been expelled from Paradise because of the disobedience of Adam and Eve. All the scientific data indicate that the golden age for man, if any, is in the future not in the past. It may be significant that in his only reported reference to that part of the Torah, Jesus quoted from the other account of creation (Genesis 1:1–2:3) when he reminded his hearers that "he who made them at the beginning made them male and female" (Matthew 19:4) although he might have referred just as readily to the wholly unscientific myth about Adam's rib (Genesis 2:21–23). More importantly, it is doubtful whether Paul's interpretation of the meaning of Jesus' death on the cross, encompassing as it did the doctrine of vicarious atonement, was in harmony with Jesus' own expectations concerning its implications for mankind.

Later in the Christian Era when churchmen began to codify the doctrinal and moral principles of what they understood to be the Christian faith, as at the Council of Nicaea in A.D. 325, many of the Pauline doctrines were incorporated in the creeds, dogmas, and liturgies of the established church. During and following the Reformation in the sixteenth century, Protestants substituted for the authority and infallibility of the Pope and the ecumenical councils the authority and infallibility of the Bible, the Holy Scriptures. Even today there are many Christians who firmly believe in the literal truth and equal value of all the documents assembled in that anthology. They hold that its historical records and its references to the physical nature of the world, even though they date from a time well before the dawn of modern scholarship, are as fundamental to Christian faith as its religious concepts.

All of which is to say that if one is to become acquainted with Christianity as the religion of Jesus Christ, he must strip away the husks of traditional beliefs and dogmatic interpretations and seek the kernel of truth in Jesus' own words and deeds. Impediments to that search are obvious; his sayings are hearsays, he conveyed his thoughts in parables, his reported words are often cryptic, and his actions were recorded by fallible observers. But the impression he made upon his associates is unmistakable, his personality comes through with crystal clarity, his basic philosophy of life is readily discernible.

Central in the religion of Jesus is his concept of God as loving heavenly Father; the spiritual aspects of the administration of the universe are kindly, merciful, just, wise, adorable. God is spirit, unrestricted by space or time (John 4:20–24). The directives of the spiritual field are of prime importance (Matthew 4:4) but they do not override the regulations of the gravitational or any other field (Matthew 4:5–7). Indeed, Jesus was quite in line with his predecessors among the Hebrew sages in insisting that God is a God of Law. I imagine that if he had been on a college campus in this twentieth century he might well have called attention to the periodic table of the elements and have added to his well-known saying, "consider the lilies of the field, how they grow" (Matthew 6:28), the equally good advice, "consider the atoms and how they change." He might also have approved the insertion, between verses

six and seven in the Nineteenth Psalm, of some such verse
as

> "Within the atom there is obedience to law,
> And the elements are ordered precisely;
> Their voice repeats the heavenly theme."[11]

The most extraordinary factor in Jesus' theology was,
however, his supreme faith that God is a God of Love,
that love is paramount among the forces operating in the
spiritual sector of the life of man. In this faith he was joined
whole-heartedly by Paul, in whose epistles some of the
most majestic passages deal with this attribute of God and
what it means for men and women. The confidence that
Jesus had in the power of love was derived not so much
from his knowledge *about* the administration of the universe as from his knowledge *of* that administration, knowledge gained when his spirit was in communion with the
Divine Spirit, especially in times of prayer and contemplation. So close was his communion with his Heavenly
Father that he could proclaim with honesty and humility:
"I and my Father are one" (John 10:30). The quality of his
life and the power of the love he displayed were such that
many of those who associated with him and many who
have come to know him in later years have considered him
to be the Son of God. It is possible that his knowledge *of*
the administration of the universe was more complete than
that of any other person in human history. Moreover, and
surprisingly enough, his teachings indicate profound
knowledge *about* the fundamental laws of nature. With
scarcely an exception his teachings ring true when compared with the modern scientific understanding of those
basic principles.[12]

With a faith so firm that he was willing to die for it,
Jesus opted for the third of the three permissible speculations concerning the autonomy of the human soul, set

forth on a preceding page. What he might have thought about the evolution of the soul, I do not know, but he surely believed that it is the privilege of human beings to cultivate the human spirit to the greatest possible extent (Mark 8:36, 37). Faith in the potential immortality of the soul as promulgated by Jesus is an essential part of the Christian religion. It has brought solace and comfort to countless Christians in times of bereavement and sorrow. As I have said before, it is completely within the domain of religious faith, susceptible neither to scientific validation nor to negation in the present state of human knowledge.

Jesus was a teacher of the Judaic scriptures (Mark 1:21, 22, *et al.*), and it was natural for him to think of sin as any violation of the traditional Judaic laws or commandments. He was, however, a noncomformist and he recognized the difference between petty and trivial regulations on one hand and fundamental and significant ones on the other (Luke 14:11–16 *et al.*). He stressed inner thoughts and motives rather than overt deeds as indicative of man's righteousness or lack thereof (Matthew 23:23–28 *et al.*). Thus he selected as the two great commandments: "Thou shalt love the Lord thy God" and "Thou shalt love thy neighbor as thyself" (Matthew 22:36–40; Mark 12:28–31); or in the parlance I am using here: "Discover, respect, and abide by the administrative directives pertaining to the spiritual aspects of human life" and "Accept responsibility for the welfare of all other persons in the same way and to the same extent as you accept it for your own welfare." Violation of those commandments is sin.

As for religion in general, so for the Christian religion in particular, virtue is not merely the absence of sin but is a dynamic commitment to positive thoughts and actions. In addition to obedience to the real essence of the Judaic commandments, to which Jesus said the scribes and Pharisees were giving only lip service, he exhorted his followers to keep *his* commandments. To Simon Peter he said:

"Feed my sheep" (John 21:15–17); to the rich ruler he said: "Sell all that thou hast, and distribute unto the poor" (Luke 18:18–24); and to all he said "Whosoever will be chief among you, let him be your servant" (Matthew 20:27). Here too the emphasis is placed upon the inner life of man, the secret thoughts and hidden desires; it is the "pure in heart" who "shall see God" (Matthew 5:8). Quite definitely, it is not easy to be a virtuous Christian.

Salvation is one of the most strenuously debated of the key words in Christian theology. Many of the reported sayings of Jesus that bear upon that aspect of his religion were in the format of parables; others were allegorical; and some appear mutually contradictory. Nevertheless certain basic principles can be identified. Among them is a considerable element of permissiveness. "Behold, I stand at the door and knock; if any man hear my voice, and open the door, I will come in to him, and will sup with him, and he with me" (Revelation 3:20). No suggestion of force or coercion; it is an opportunity available to all, to be taken or rejected as each decides. Or again, in the parable about wheat and tares (Matthew 13:24–30), the tares are permitted to grow until the harvest which is described as a time of judgment, a word that Jesus used on many occasions and in diverse connections. I am reminded of the processes of geologic life development in which genetic mutations and variations in gene quotas produce creatures having new capabilities. Such creatures are permitted to test the survival value of their novel behavior in the continuing competition for existence; but sooner or later the processes of natural selection pass judgment upon them, approving some and rejecting others. He who finds himself at one in spirit with Jesus, and therefore at one with the administration of the universe, is blest by all the salvation that humankind can expect. Salvation, thus conceived, is a result of both works and faith. The follower of Jesus must at least open the door; after that there is much work to be done as he strives to keep Jesus' commandments. There is also faith in the tender mercy and abiding love of the Heavenly Father to whom Jesus taught his followers to pray and whose character he was making known. It is more in keeping with Jesus' own conviction about the meaning of his life to say that he *lived* to save men's souls than to say that he *died* to save men's souls.

Destiny is a word that rarely appears in the several current translations of the sayings attributed to Jesus, but his concept of the destiny of human beings is abundantly clear: become a worthy citizen of the Kingdom of God or of the Kingdom of Heaven, apparently a synonym for the same concept. Central in Jesus' mission in life was the proclamation of the good news, the gospel, that "the kingdom of heaven is at hand" (Matthew 4:17). The kingdom "is not of this world (John 18:36); it is "within you" (Luke 17:21); it is the realm of the spirit. In the secular vocabulary I use when speaking as a scientist *per se*, it is the postulated spiritual field or, more generally, the spiritual aspect of the administration of the universe. As glimpsed by certain Hebrew prophets, such as Daniel (4:3), long before the birth of Jesus, it is an everlasting kingdom.

We will never know precisely what happened to the body of Jesus after his death on the cross. The several reports that he was seen in the flesh may have been derived from hallucinatory experiences, as Preston Harold interprets them.[13] One thing we do know, however. After his crucifixion his little band of trusting companions started homeward, woefully disappointed, completely disheartened, their spirits so broken that some of them feared to acknowledge that they had ever known him. Then something happened that abruptly transformed them into a courageous company whose evangelical zeal and missionary enterprise changed the course of human history. They felt the presence of his Spirit in their midst; they knew that he was with them as he had promised, "even unto the end of the world" (Matthew 28:20). Their trust in the man with whom they had been so intimate for months or years was renewed and strengthened, their belief in him as the Son of God was restored and intensified.

NOTES

1. Henry Margenau, "Tachyons," *Main Currents in Modern Thought* 26 (1969): 56–59.

2. cf. Eugene Marais, reported by Robert Ardrey in *African Genesis* (New York: Atheneum, 1961), 79–81.

3. Kirtley F. Mather, *The Evolution of the Soul*, pamphlet published by William F. Ayres Foundation, Plymouth Congregational Church, Lansing, Michigan, 1940; Edmund W. Sinnott, *The Biology of Spirit* (Viking, 1955).

4. cf. Ian C. Barbour, *Issues in Science and Religion* (New York: Prentice-Hall, 1966), 359–63.

5. cf. Archie J. Bahm, *The World's Living Religions* (New York: Dell Laurel Edition, 1964).

6. Henry Van Dyke, "Hudson's Last Voyage," *The Poems of Henry Van Dyke* (New York: Charles Scribner's Sons, 1920).

7. Robert Ulich, editor, *Education and the Idea of Mankind* (New York: Harcourt, Brace and World, 1964).

8. Pierre Teilhard de Chardin, *The Phenomenon of Man* (New York: Harper and Brothers, 1959), 180–84, 272–76, *et al.*

9. cf. Gardner Murphy and Laura A. Dale, editors, *Challenge of Psychical Research* (New York: Harper, 1961).

10. Kirtley F. Mather, "Creation and Evolution," *Science Ponders Religion*, ed. Harlow Shapley (New York: Appleton-Century-Crofts, 1960), chapter 3, 32–45; "Geology and Genesis," *Main Currents in Modern Thought* 21 (1964): 10–16; also published as a chapter in *D-days at Dayton: Reflections on the Scopes Trial*, ed. Jerry R. Tompkins (Baton Rouge: Louisiana State University Press, 1965).

11. Kirtley F. Mather, *The World in Which We Live* (Philadelphia: Education Press and Boston: Pilgrim Press, 1961), Teacher's edition, 32.

12. Preston Harold, *The Shining Stranger* (New York: The Wayfarer Press, Dodd, Mead, and Co., 1967), 199–213.

13. Preston Harold, *The Shining Stranger*, 311–20.

Epilogue

WE ARE LIVING IN a permissive universe. The permissiveness is, however, limited by administrative regulations and restricted by administrative directives. Law-abiding processes of change have operated in the past and continue to operate today in certain discernible directions. The over-all direction is toward orderly organization; the universe is a cosmos, not a chaos. Elementary particles of matter and quanta of energy are organized into atoms; atoms are organized to produce molecules. Some of the molecules are organized to form crystals, others to form cells. Some of the cells are organized as plants, others as animals. Some of the multicellular animals are organized to form colonies, such as a colony of corals; others to form societies, such as a society of social insects or of various species of mammals.

Human societies are the last to appear in the geologic record of organic evolution. They are composed of the most complexly organized among all creatures, with the greatest mental capacities. They are held together fundamentally by intangible forces that cannot be evaluated in scientific terms of velocity or dimension but can be evaluated in humanistic terms of their spiritual effects. In the

191

evolution of life toward mankind, and in the hundreds of thousands of years of human history, there has been a progressive increase in awareness of both the tangible and intangible factors in man's environment. This has resulted in what we call the cultural evolution of mankind, as contrasted with biological or anatomical evolution. By and large, the survival values of the various experiments now being tried in the diversified cultures of differing human societies will determine the future of man as an inhabitant of the earth.

Projecting into the future the major trend of human cultural evolution, as it has proceeded during the fairly recent past, we may envision the goal which mankind should strive to reach. That goal may be described as an exquisitely functioning social organization composed of individuals possessing such physical, mental, and spiritual attributes that each, of his own free will and accord, cooperates intelligently and wholeheartedly with all the others in using the rich resources of the bountiful earth for the continuing welfare of all mankind. To achieve that goal may well be a prime purpose of life. Indeed, it may even be thought of as a purpose of the administration of the universe, with its flair for orderly organization.

The evolution of mankind thus far has obviously been in accord with administrative regulations and directives; otherwise man would not have attained his present high estate in the animal kingdom. This is not to say, however, that man's future is assured by the cosmic administration, whether we are thinking of the destiny of individual human souls or of that of the species, *Homo sapiens*, as a whole. At no time throughout the long history of life on earth, have the processes of evolution guaranteed success for any creature under their sway; they have only guaranteed opportunities to succeed. At the moment, man has an opportunity that is unique among all creatures and in all places, past or present, known to us. Whether he will grasp that opportunity or reject it will be determined by human beings themselves. To grasp it, all of the high idealism and valid morality of religion, as well as all of the tested knowledge and prudent pragmatism of science, will be required.

Surveying the world of humanity around us, one may question whether the available human resources are adequate to meet the challenge of the opportunity now at hand. There seem to be plenty of intellectual resources, but are the spiritual resources sufficient for the task ahead? Is there enough love or too much hate? Is there such an abundance of selfishness that good will toward others is hopelessly submerged? Are we not polluting the atmosphere, the hydrosphere, and the lithosphere with almost complete disregard for our own welfare, not to mention that of generations soon to come? Does the human sense of social responsibility or stewardship have any chance of overcoming the instincts of rugged individualism inherited from our prehuman ancestors? In this time of potential abundance and inescapable interdependence, with rapidly increasing populations, are we not fragmenting by nationalistic ambitions, racial prejudices, and economic imbalances the one world toward which all the administrative directives point?

These are valid questions and the answers seem discouraging. We must remember, however, that in the record of geologic life development the processes of biological evolution have repeatedly produced from quite unpromising raw material surprisingly good results; from the lungfish of mid-Paleozoic time came the quadrupedal amphibians of late-Paleozoic time, for example. Presumably the processes of cultural evolution may operate in a similar way. In any event, we have no choice; we must do the best we can with what we are.

In the permissive universe the future is open wide for mankind, but there are regulations that must be obeyed and directives that must be accepted. In the last analysis, men and women will determine, individually and collectively, what mankind's future will be.

TIME UNITS			OROGENY	DATES	LIFE RECORD
PHANEROZOIC EON	CENOZOIC ERA	CENOZOIC PERIOD	TERTIARY PERIOD	Pleistocene epoch	Cascadian revolution	2,000,000 B.C.
			Pliocene epoch		
			Miocene epoch		
			Oligocene epoch		
			Eocene epoch		
			Paleocene epoch		70,000,000 B.C.
	MESOZOIC ERA		CRETACEOUS PERIOD	Laramide revolution	
			JURASSIC PERIOD	Nevadian disturbance or revolution	170,000,000 B.C.
			TRIASSIC PERIOD	Palisade disturbance	
	PALEOZOIC ERA		PERMIAN PERIOD	Appalachian revolution	200,000,000 B.C.
		CARBONIFEROUS	PENNSYLVANIAN PERIOD		
			MISSISSIPPIAN PERIOD		
			DEVONIAN PERIOD	Acadian disturbance	260,000,000 B.C.
			SILURIAN PERIOD		
			ORDOVICIAN PERIOD	Taconian disturbance	350,000,000 B.C. / 400,000,000 B.C.
			CAMBRIAN PERIOD		500,000,000 B.C.
CRYPTOZOIC EON	PRE-CAMBRIAN	(Only local sequences can be recognized)	LATE PRE-CAMBRIAN (PROTEROZOIC)	Penokean orogeny	
			EARLY PRE-CAMBRIAN (ARCHEOZOIC)	Algoman revolution / Laurentian revolution	2,000,000,000 B.C.

Life Record labels: Man, Carnivores, Camels, Elephants, Whales, Bats, Apes, Horses, Monkeys, Birds, Marsupials, Turtles, Insectivores, Crocodiles, Angiosperms, Grasses, Mesozoic mammals, Dinosaurs, Pterosaurs, Plesiosaurs, Toothed birds, Ichthyosaurs, Conifers, Cycads, Theriodonts, Ammonites, Cotylosaurs, Insects, Seed ferns, Crinoids, Labyrinthodont, Sharks, Clams, Cordaites, Blastoids, Choanichthyes, Starfish, Brachiopods, Ferns, Scale trees, Corals, Cystoids, Jawless fishes, Nautiloid, Snails, Trilobites, Sponges, Algae

Appendix 1. Geologic Time Chart. Ascending lines indicate the range in time of most of the chief groups of animals and plants. If the line ends in a crossbar, this denotes the time of extinction; if it ends in a dart, the group is still living. (Reproduced by permission from Carl O. Dunbar and Karl M. Wage, *Historical Geology*, © 1949, John Wiley & Sons, Inc.)

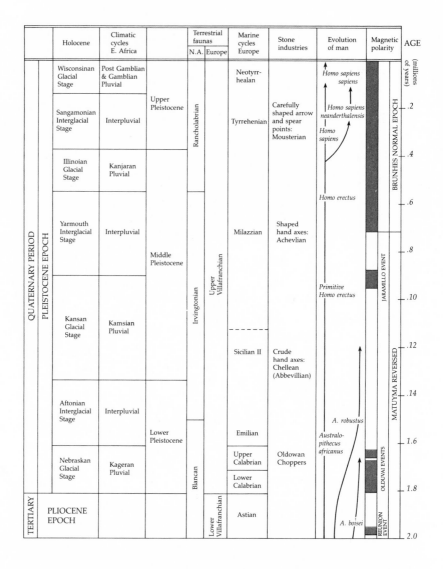

Appendix 2. A provisional stratigraphic chart comparing climatic cycles, marine cycles, stone industries, and the ranges of the ancestors of man. (Reproduced by permission of Harper & Row from Carl K. Seyfert and Leslie A. Sirkin, *Earth History and Plate Tectonics*, 2nd Edition, © 1979, Carl K. Seyfert and Leslie A.Sirkin.)

Appendix 3
The Geologist's Timetable

Dates (m.y.a.) indicate millions of years ago
and are approximate.

Cenozoic Era   65 m.y.a. to present time
    The era of modern life, during which the earth's ani-
    mals and plants attained their present forms and dis-
    tributions and the continents and ocean basins came
    to have their existing topographies and positions.

QUATERNARY PERIOD   1.5 m.y.a. to present time
    The second and shorter of the two periods which com-
    prise the Cenozoic Era. The name Quaternary no longer
    has any chronologic significance.

Recent or Holocene Epoch   0.01 m.y.a. to present
    At and near localities subjected to glaciation by conti-
    nental ice sheets during the preceding epoch, all events
    occurring since the disappearance of that ice are usually
    referred to this epoch. Dating it as the last ten thousand
    years is arbitrary, but convenient. By this time, organic
    evolution had brought a wholly modern (holocene) ap-
    pearance to animals and plants, and during it the cul-
    tural evolution of mankind became the most crucial
    factor in terrestrial life.

Pleistocene Epoch   1.5 to 0.01 m.y.a.
    The Great Ice Age. A time of widespread ice sheets in

197

northern North America, northwestern Europe, northern Asia, southern South America, and Antarctica, with
extensive glaciers in lofty mountains the world around.
Animals and plants were almost modern (pleistocene),
but not quite completely so. During this time, mankind
spread widely over all continents except Antarctica and
deployed into several species and varieties, some of
which are now extinct.

TERTIARY PERIOD   65 to 1.5 m.y.a.
The Age of Mammals, during which the continents
gradually came to have approximately their present positions, and the ocean-basin floors spread laterally to
about their present dimensions. The great mountain
chains ringing the Pacific Ocean and crossing southern
Europe and Asia attained nearly their present altitudes
as a result of crustal movements and volcanic activity.
The name Tertiary no longer has any chronologic significance.

Pliocene Epoch   13 to 1.5 m.y.a.
The time when animals and plants became more modern (pliocene) in their appearance than ever before and
subhuman hominids became clearly differentiated from
ancestral apes.

Miocene Epoch   25 to 13 m.y.a.
The time when animals and plants attained a somewhat
modern (miocene) appearance. Horses, cattle, camels,
and elephants were the dominant mammals of the land
and were much more widespread than today. Several
kinds of anthropoids flourished in Africa, Europe, and
Asia, and grasses and cereals first became a large component of the land vegetation.

Oligocene Epoch   36 to 25 m.y.a.
The time when animals and plants had a slightly modern (oligocene) appearance. All the mammalian families
living today, except the hominidae, were represented
in the fossil record, but most of the fossils indicate creatures intermediate between the archaic forms of the
Eocene Epoch and the more modern forms of the Miocene Epoch. Anthropoid apes and primitive elephants
first appear in the geologic record.

Eocene Epoch 58 to 36 m.y.a.

The name means dawn of the recent and refers to the presence among invertebrate fossils of some that can be assigned to species living today. Many of the mammals belonged to orders that became extinct by the end of the epoch; others, like Eohippus, the dawn horse, were decidedly primitive in comparison with their modern descendants. The order known as Primates, which eventually included man, was represented by creatures resembling modern lemurs and tarsiers.

Paleocene Epoch 65 to 58 m.y.a.

During the epoch, the ancient dawn of the recent, placental mammals spread throughout Eurasia, Africa, and both Americas.

Mesozoic Era 230 to 65 m.y.a.

The name indicates the intermediate nature of the life of this time, between the ancient faunas and floras of the Paleozoic Era and the more modern life of the Cenozoic Era. It is often called The Age of Reptiles, inasmuch as that class of vertebrates dominated land, sea, and air throughout much of this interval of time.

CRETACEOUS PERIOD 135 to 65 m.y.a.

The name derives from the presence of thick and extensive beds of chalk among the sedimentary rocks formed during this time, as in the White Cliffs of Dover. The close of the period is marked by the extinction of the dinosaurs, the flying reptiles, and the aquatic reptiles that had earlier reached the summit of their long career. Notable also is the advent of placental mammals in Asia not long after the middle of the period.

JURASSIC PERIOD 180 to 135 m.y.a.

Named after the Jura Mountains in Switzerland. Its predominantly sedimentary rocks contain fossils of the first birds, but flying reptiles were the masters of the air. Egg-laying and marsupial mammals were widely distributed. Although ferns and evergreens were still the most abundant plants in the land, vegetation, deciduous trees and flowering plants were becoming more abundant.

TRIASSIC PERIOD   230 to 180 m.y.a.

Named for the three-fold sequence of sedimentary rocks
in central Germany to which the term was applied in
1834. Its marine invertebrate faunas are strikingly different from those of late Paleozoic time. There was,
however, no sudden change in the continuing development of amphibians, reptiles, and land vegetation.
Transitional forms from certain reptiles to primitive
mammals are recorded at several localities in Africa and
North America.

Paleozoic Era   600 to 230 m.y.a.

The era of ancient life. In contrast to the rocks of Precambrian time, the abundant fossils in its sedimentary
rocks facilitate the correlation of events from place to
place the world around. The latter part of the era is
sometimes referred to as the age of fishes and the earlier
part as the age of invertebrates.

PERMIAN PERIOD   280 to 230 m.y.a.

Named for the Permian Basin, west of the Ural Mountains in the U.S.S.R. A time of widespread deserts and
more-than-usual crustal deformation, it was also a time
when ice sheets spread over vast areas in Asia, Africa,
Australia, Antarctica, and South America. By its close,
many previously flourishing families and orders of invertebrate animals had become extinct. During it, several of the modern orders of insects emerged and the
reptiles deployed into new orders; some of these were
short-lived, but others included distinctly mammal-like
forms, as well as the ancestors of the Mesozoic reptiles.

PENNSYLVANIAN OR UPPER CARBONIFEROUS PERIOD   305 to 280
m.y.a.

The names derived (1) from Pennsylvania, where rocks
of this age are widely exposed, and (2) from the system
of rocks in Great Britain, Germany, and the Soviet Union
that includes the Coal Measures and therefore is rich
in carbon. The lush vegetation of the extensive coal-
forming swamps and marshes consists largely of ferns
and nonflowering plants (gymnosperms) although
flowering plants (angiosperms) are first recorded among
its fossils. The transition from amphibians to reptiles

was accomplished during this period. Permo-Carbon-
iferous glaciation may have affected some areas toward
its close, although most of that ice age is definitely dated
as Permian.

MISSISSIPPIAN OR LOWER CARBONIFEROUS PERIOD   345 to 305
m.y.a.

The names are derived (1) from the Mississippi Valley
and (2) from the lower part of the system of European
rocks, the upper part of which is correlated with the
Pennsylvanian system of North America. During it, the
amphibians became well established as partly aquatic
and partly land-dwelling vertebrates.

DEVONIAN PERIOD   400 to 345 m.y.a.

Named for Devonshire, southwestern England, where
rocks of this age were studied early in the nineteenth
century. Among the fossils found in its sedimentary
rocks are those of the earliest known amphibians. Much
more numerous are the fossils of various kinds of sharks
and fringed-fin fishes.

SILURIAN PERIOD   425 to 400 m.y.a.

Named for an ancient Celtic tribe who once lived in the
part of Wales where rocks of this age were studied early
in the nineteenth century. Among the animals of this
time were scorpions, quite similar to modern forms and
probably the first air-breathing animals; primitive fish
are the only known vertebrates of this period.

ORDOVICIAN PERIOD   500 to 425 m.y.a.

Named for an aboriginal tribe who once lived in the
part of Wales where the sedimentary rocks were first
assigned to this period in 1879. In general, its strata
contain a greater abundance and variety of invertebrate
fossils than do the Cambrian strata. Among its fossils
are those of primitive fish-like animals, the oldest known
vertebrates.

CAMBRIAN PERIOD   600 to 500 m.y.a.

The name perpetuates an early designation for Wales,
Cambria. Its strata are the earliest to contain an abundant fossil record of the contemporary life. Among its
fossils are representatives of every major group or phy-

lum of invertebrate animals; no vertebrate fossils are known. Plant life is recorded by fossil algae and the spores of moss-like vegetation; no vascular plants are known.

Precambrian Time   From earth's origin about 4,500 m.y.a. to
600 m.y.a.

Some geologists designate the latter part of this large fraction of earth history as the Protozoic Era and an earlier part as the Archeozoic Era but there is no consensus concerning the dating of the milestone that would separate those two eras. The oldest known rocks were formed about 3,500 m.y.a. Most of the rocks formed during this ancient time have been subjected to such high temperatures and great pressures that they could not possibly have preserved any fossil record of contemporary life. The unmetamorphosed or only slightly metamorphosed sedimentary rocks contain extremely rare fossils of only one-celled or worm-like animals and one-celled or moss-like plants. Some of these date back to upwards of 3,000 m.y.a.

Afterword
by
Dr. Archie Bahm

THIS BOOK IS a prominent geologist's philosophy of life—the life of the universe's several billion-year-long progressive organic evolution toward an ideal human society as a goal.

His universe of law and order permits progress through trial-and-error experimentation, survival of what fits, and progress through more highly complex kinds of organization—chemical, organic, and cultural. Such progress, which is not guaranteed, but permitted, can be observed. Dr. Mather has interestingly summarized multi-stage developments with a rich plethora of relevant samples from many scientific fields. He proposes a naturalistic reconciliation of science and religion by reinterpreting religion in ways consistent with his enormous store of scientific, primarily geological, knowledge.

One distinctive contribution is his postulating with assurance an administrative order in the universe providing both regulating conditions (deterministic laws) and guiding directives (permitting evolution of self-organizing systems) that serve as bases for thinking of the universe as a whole as a spiritual force that has been interpreted by Christians as God. His God, however, is both impersonal and internal to the universe, not an arbitrary, willful, supernatural person.

Professor Mather has exercised considerable leadership in liberalizing intellectual attitudes, and he has popularized geological information to provide a more solid scientific foundation that needs to be taken into account by liberal religious leaders.

This work is the finished product of a lifetime of thinking by a recognized leader and emphasizes the significance of geological factors for sound religious belief. His work is one of the best representatives of the intellectually liberalizing influences of scientific developments on religious beliefs in recent decades. I must agree with the author that the audience addressed is "broadly inclusive and widely diversified." His clear and interesting accounts of geological developments are likely to have more appeal to nongeologists than to geologists already familiar with the facts. He speaks a language understandable by any middle American concerned with reconciling science and religion and willing to take account of the sometimes disturbing scientific facts.

Dr. Archie Bahm, Professor of Philosophy, Emeritus, the University of New Mexico 1973, received his Ph.D. at the University of Michigan in 1933. He was Phi Beta Kappa scholar, Albion College 1929, Fulbright Research Scholar University of Rangoon, 1955–56, and is author of 22 books on Philosophy, Ethics, and Religion.

Afterword
by
Dr. Donald H. Rhoades

THIS BOOK IS THE summary, inclusive statement of the educator, concerned more for the life-value for the reader of what is said than for the academic content of the perspectives he tries to share. Indeed, perspective is the key issue, the kind of perspective that can frame and undergird one's sense of place in a felt wholeness of the significantly real, a philosophy by which to live, and to give intellectual support for being "at home" in one's total context. Because it was written so late in life, *The Permissive Universe* may well be read as the last contribution of a long generation of scientists who reached out beyond their specialties to present them as parts of a world-view. Here, Kirtley Mather adds his contribution to those of Margenau, Shapley, Sinnott, Dobzhanski, Dubos, and, perhaps especially, Whitehead and T. de Chardin. In all of them there was a movement beyond narrow specialization, beyond conflicts of imageries and vocabularies, toward integration and wholeness.

For Mather, our time of trouble is definitely *a* time of trouble, new in its particulars, but taking its place in the long succession of times of change, crisis, danger and opportunity, failure and progress, that have marked the evolutionary story. And, because the human stage is one of

reflection, the appropriate response to troubles is reexamination, rather than Pollyanaism or despair.

Our troubles are, most broadly, population, rising expectations, and knowledge, with knowledge (particularly
medicine and communication) as the *sine qua non* of the
other two. We have been "catapulted into a new age." The
author nowhere refers to the "thrownness" (Einworfen-
heit) bewailed by some existentialists, perhaps because the
imagery wasn't very new to him, and because dread-rid-
den introspectiveness was so foreign to his own nature.
Our immediate concerns are factors in social evolution,
"extraordinarily swift," but, because they are social, they
are in principle manageable. This increasingly self-direct-
ing phase of evolution "has just begun."

The author might well have headed a chapter with L.
P. Jacks' statement of some years ago: "Man is organic to
the universe"; only the wording is missing. If man must
be understood in terms of his total context (past, present,
and future), the reverse is also true. The universe is some
kind of unity, revealing, or understandable in terms of,
regulations, guide-lines, and varying degrees of what Dr.
Mather chooses to call "permissiveness," most importantly
openness to human choice and effective initiative. In this
totality, what we call the qualitative is equally as real as
what we call the quantitative. Ethical and esthetic values
emerge, following life, as integral elements in evolution
as a whole, continuous with and growing out of the *world*
of the natural sciences. But growing out of does not mean,
a priori, entirely caused by.

The viewpoint here presented finds it necessary and
justified to speak of "some kind of administration." This
is no appeal to some traditional anthropomorphic administrator, but a suggestive positing of a field of spiritual
force, not merely analogous to, but significantly comparable to, the fields of gravitational and magnetic forces. To
those who might object, or fear that these affirmations go

beyond valid knowledge to baseless speculation, the study points out that both *scientific* and *spiritual* judgments rely on both perceptual and conceptual factors, and that the *scientific* has come to mean, broadly, measuring and the measureable in terms of space and time. And since most of what men live by and for is not amenable to calibration, there is no reason for any truly comprehensive approach to reality to accept a *scientific* veto with respect to matters outside the scope of scientific method. Does all this justify "religion"? Yes, if religion is the general term for attitudes, images, and practices that bring wholeness to the life of individuals and groups. But it finds no place for imperialistic dogmatism.

As a humanist-scientist, Dr. Mather is naturally an ecologist. As Anteus could maintain strength for his struggle only by immediate contact with his mother Earth (Gea), so must mankind rely on and manage competently and even faithfully the resources of planet earth. Here the author is (still) the qualified optimist. Recognizing that some statistics have changed since *Enough and to Spare,* he finds the plusses and minuses more or less keeping pace, and, despite the projections of the Club of Rome, affirms that "the future is still wide open for mankind" on an earth habitable for millions of years to come. Changes will occur, of course, including ice ages, but they should not be fatal. Man could commit nuclear suicide, but, as of 1975, the author does not see that as likely. *Homo sapiens* has a largely unique opportunity to go on moving ahead, in a dependable, supportive, and permissive universe.

Donald H. Rhoades, Professor Theology and Philosophy, emeritus, the University of Southern California and Claremont Graduate School, received his Ph.D. at Yale University. He is past president of the California Conference on Science and Religion. He received his Bachelor's degree as a geology major at Colby College.

Books
by Kirtley F. Mather

Science in Search of God, Henry Holt and Company, 1928

Old Mother Earth, Harvard University Press, 1928

Sons of the Earth, W. W. Norton and Co., 1930

Laboratory Manual of Physical and Historical Geology (with C. J. Roy), D. Appleton-Century Company, 1934

Adult Education: A Dynamic for Democracy (with Dorothy Hewitt), D. Appleton-Century Company, 1937

A Source Book in Geology (with Shirley L. Mason), McGraw-Hill, 1939

Enough and to Spare, Harper and Brothers, 1944

Crusade for Life (John Calvin McNair Lectures), University of North Carolina Press, 1949

A Laboratory Manual for Geology: I, Physical Geology (with Chalmer J. Roy and Lincoln R. Thiesmeyer), Appleton-Century-Crofts, 1950

A Laboratory Manual for Geology: II, Historical Geology (with C. J. Roy) Appleton-Century-Crofts, 1952

The World in Which We Live ("A Course for Older Young People and Adults"), Christian Education Press, Phila., and Pilgrim Press, Boston, 1961

The Earth Beneath Us, Random House, NY, 1964; revised edition, 1975

Source Book in Geology, 1900–1950, Harvard University Press, 1967

Index